ASE Test Preparation Series

Medium/Heavy Duty Truck Test

Preventive Maintenance (Test T8)

4th Edition

THOMSON

DELMAR LEARNING

Australia   Canada   Mexico   Singapore   Spain   United Kingdom   United States

THOMSON

DELMAR LEARNING

Thomson Delmar Learning's ASE Test Preparation Series
Medium/Heavy Duty Truck Test for Preventive Maintenance (Test T8), 4th Edition

Vice President, Technology Professional Business Unit:
Gregory L. Clayton

Product Development Manager:
Kristen L. Davis

Product Manager:
Kimberley Blakey

Editorial Assistant:
Vanessa Carlson

Director of Marketing:
Beth A. Lutz

Marketing Manager:
Brian McGrath

Marketing Coordinator:
Jennifer Stall

Production Director:
Patty Stephan

Production Manager:
Andrew Crouth

Content Project Manager:
Kara A. DiCaterino

Art Director:
Robert Plante

Cover Design:
Michael Egan

ISBN: 1-4180-4836-4

NOTICE TO THE READER

Publisher does not warrant or guarantee any of the products described herein or perform any independent analysis in connection with any of the product information contained herein. Publisher does not assume, and expressly disclaims, any obligation to obtain and include information other than that provided to it by the manufacturer.

The reader is expressly warned to consider and adopt all safety precautions that might be indicated by the activities herein and to avoid all potential hazards. By following the instructions contained herein, the reader willingly assumes all risks in connection with such instructions.

The publisher makes no representation or warranties of any kind, including but not limited to, the warranties of fitness for particular purpose or merchantability, nor are any such representations implied with respect to the material set forth herein, and the publisher takes no responsibility with respect to such material. The publisher shall not be liable for any special, consequential, or exemplary damages resulting, in whole or part, from the readers' use of, or reliance upon, this material.

Contents

Section 5 Sample Test for Practice

Section 6 Additional Test Questions for Practice

Section 7 Appendices

Preface

Delmar Learning is very pleased that you have chosen our ASE Test Preparation Series to prepare yourself for the ASE Medium/Heavy Truck Examination. These guides are available for all of the medium/heavy truck areas including T1-T8. These guides are designed to introduce you to the Task List for the test you are preparing to take, give you an understanding of what you are expected to be able to do in each task, and take you through sample test questions formatted in the same way the ASE tests are structured.

If you have a basic working knowledge of the discipline you are testing for, you will find Delmar Learning's ASE Test Preparation Series to be an excellent way to understand the "must know" items to pass the test. These books are not textbooks. Their objective is to prepare the technician who has the requisite experience and schooling to challenge ASE testing. It cannot replace the hands-on experience or the theoretical knowledge required by ASE to master vehicle repair technology. If you are unable to understand more than a few of the questions and their explanations in this book, it could be that you require either more shop-floor experience or further study. Some resources that can assist you with further study are listed on the rear cover of this book.

Each book begins with an item-by-item overview of the ASE Task List with explanations of the minimum knowledge you must possess to answer questions related to the task. Following that there are two sets of sample questions followed by an answer key to each test and an explanation of the answers to each question. A few of the questions are not strictly ASE format but were included because they help teach a critical concept that will appear on the test. We suggest that you read the complete Task List Overview before taking the first sample test. After taking the first test, score yourself and read the explanation to any questions that you were not sure about, including the questions you answered correctly. Each test question has a reference back to the related task or tasks that it covers. This will help you to go back and read over any area of the task list that you are having trouble with. Once you are satisfied that you have all of your questions answered from the first sample test, take the additional tests and check them. If you pass these tests, you will be prepared to do well on the ASE test.

Our Commitment to Excellence

The 4th edition of Delmar Learning's ASE Test Preparation Series has been through a major revision with extensive updates to the ASE's task lists, test questions, and answers and explanations. Delmar Learning has sought out the best technicians in the country to help with the updating and revision of each of the books in the series.

To promote consistency throughout the series, a series advisor took on the task of reading, editing, and helping each of our experts give each book the highest level of accuracy possible.

Thank you for choosing Delmar Learning's ASE Test Preparation Series. All of the writers, editors, Delmar Staff have worked very hard to make this series second to none. It is our objective to constantly improve our product at Delmar by responding to feedback.

If you have any questions concerning the books in this series, email us at: autoexpert@trainingbay.com.

1   The History and Purpose of ASE

ASE began as the National Institute for Automotive Service Excellence (NIASE). It was founded as a non-profit independent entity in 1972 by a group of industry leaders with the single goal of providing a means for consumers to distinguish between incompetent and competent technicians. It accomplishes this goal by testing and certification of repair and service professionals. From this beginning it has evolved to be known simply as ASE (Automotive Service Excellence) and today offers more than 40 certification exams in automotive, medium/heavy duty truck, collision, engine machinist, school bus, parts specialist, automobile service consultant, and other industry-related areas. At this time there are more than 400,000 professionals with current ASE certifications. These professionals are employed by new car and truck dealerships, independent garages, fleets, service stations, franchised service facilities, and more. ASE continues its mission by also providing information that helps consumers identify repair facilities that employ certified professionals through its Blue Seal of Excellence Recognition Program. Shops that have a minimum of 75 percent of their repair technicians ASE certified and meet other criteria can apply for and receive the Blue Seal of Excellence Recognition from ASE.

ASE recognized that educational programs serving the service and repair industry also needed a way to be recognized as having the faculty, facilities, and equipment to provide a quality education to students wanting to become service professionals. Through the combined efforts of ASE, industry, and education leaders, the non-profit National Automotive Technicians Education Foundation (NATEF) was created to evaluate and recognize training programs. Today more than 2,000 programs are ASE certified under the standards set by the service industry. ASE/NATEF also has a certification of industry (factory) training program known as CASE. CASE stands for Continuing Automotive Service Education and recognizes training provided by replacement parts manufacturers as well as vehicle manufacturers.

ASE certification testing is administered by the American College Testing (ACT). Strict standards of security and supervision at the test centers ensure that the technician who holds the certification earned it. Additionally, ASE certification also requires that the person passing the test be able to demonstrate that they have two years of work experience in the field before they can be certified. Test questions are developed by industry experts that are actually working in the field being tested. There is more detail on how the test is developed and administered in the next section. Paper-and-pencil tests are administered twice a year at over 700 locations in the United States. Computer-based testing is now also available with the benefit of instant test results at certain established test centers. The certification is valid for five years and can be recertified by retesting. So that consumers can recognize certified technicians, ASE issues a jacket patch, certificate, and wallet card to certified technicians and makes signs available to facilities that employ ASE certified technicians.

You can contact ASE at any of the following:

National Institute for Automotive Service Excellence
101 Blue Seal Drive S.E.
Suite 101
Leesburg, VA 20175
Telephone 703-669-6600
FAX 703-669-6123
www.ase.com

WE SUPPORT
VOLUNTARY TECHNICIAN
CERTIFICATION THROUGH

National Institute for
AUTOMOTIVE
SERVICE
EXCELLENCE

2 Take and Pass Every ASE Test

Participating in an Automotive Service Excellence (ASE) voluntary certification program gives you a chance to show your customers that you have the "know-how" needed to work on today's modern vehicles. The ASE certification tests allow you to compare your skills and knowledge to the automotive service industry's standards for each specialty area.

If you are the "average" automotive technician taking this test, you are in your mid-30s and have not attended school for about 15 years. That means you probably have not taken a test in many years. Some of you, on the other hand, have attended college or taken postsecondary education courses and may be more familiar with taking tests and with test-taking strategies. There is, however, a difference in the ASE test you are preparing to take and the educational tests you may be accustomed to.

How are the tests administered?

ASE test are administered at over 750 test sites in local communities. Paper-and-pencil tests are the type most widely available to technicians. Each tester is given a booklet containing questions with charts and diagrams where required. You can mark in this test booklet, but no information entered in the booklet is scored. Answers are recorded on a separate answer sheet. You will enter your answers, using a number 2 pencil only. ASE recommends you bring four sharpened number 2 pencils that have erasers. Answer choices are recorded by coloring in the blocks on the answer sheet. The answer sheets are scanned electronically and the answers tabulated. For test security, test booklets include randomly generated questions. Your answer key must be matched to the proper booklet so it is important to correctly enter the booklet serial number on the answer sheet. All instructions are printed on the test materials and should be followed carefully.

ASE has introduced Computer-based testing (CBT) at some locations. While the test content is the same for both testing methods the CBT tests have some unique requirements and advantages. It is strongly recommended that technicians considering the CBT tests go the ASE web page at www.ase.com and review the conditions and requirements for this type of test. There is a demonstration of a CBT that allows you to experience this type of test before you register. Some technicians find this style of testing provides an advantage, while others find operating the computer a distraction. One significant benefit of CBT is the availability of instant results. You can receive your test results before you leave the test center. CBT also offers increased flexibility in scheduling. The cost for taking CBTs is slightly higher than paper-and-pencil tests and the number of testing sites is limited. The first-time test taker may be more comfortable with the paper-and-pencil tests but technicians now have a choice.

Who writes the questions?

The questions are written by service industry experts in the area being tested. Each area will have its own technical experts. Questions are entirely job related. They are designed to test the skills you need to be a successful technician. Theoretical knowledge is important and necessary to answer the questions, but the ability to apply that knowledge is the basis of ASE test questions.

Each question has its roots in an ASE "item-writing" workshop where service representatives from automobile manufacturers (domestic and import), aftermarket parts and equipment manufacturers, working technicians, and vocational educators meet in a workshop setting to share ideas and translate them into test questions. Each test question written by these experts must survive review by all members of the group.

The questions are written to deal with practical application of soft skills and system knowledge experienced by technicians in their day-to-day work.

All questions are pre-tested and quality-checked on a national sample of technicians. Those questions that meet ASE standards of quality and accuracy are included in the scored sections of the tests; the "rejects" are sent back to the drawing board or discarded altogether.

Each certification test is made up of between 40 and 80 multiple choice questions.

Note: Each test could contain additional questions that are included for statistical research purposes only. Your answers to these questions will not affect your score, but since you do not know which ones they are, you should answer all questions on the test. The five-year Recertification Test will cover the same content areas as those previously listed. However, the number of questions in each content area of the Recertification Test will be reduced by about one-half.

Objective Tests

A test is called an objective test if the same standards and conditions apply to everyone taking the test and there is only one correct answer to each question.

Objective tests primarily measure your ability to recall information. A well-designed objective test can also test your ability to understand, analyze, interpret, and apply your knowledge. Objective tests include true-false, multiple choice, fill in the blank, and matching questions. ASE's tests consist exclusively of four-part multiple choice objective questions.

The following are some strategies that may be applied to your tests.

Before taking an objective test, quickly look over the test to determine the number of questions, but do not try to read through all of the questions. In an ASE test, there are usually between 40 and 80 questions, depending on the subject. Read through each question before marking your answer. Answer the questions in the order they appear on the test. Leave the questions blank that you are not sure of and move on to the next question. You can return to those unanswered questions after you have finished the others. They may be easier to answer at a later time after your mind has had additional time to consider them on a subconscious level. In addition, you might find information in other questions that will help you recall the answers to some of them.

Do not be obsessed by the apparent pattern of responses. For example, do not be influenced by a pattern like **D, C, B, A, D, C, B, A** on an ASE test.

There is also a lot of folk wisdom about taking objective tests. For example, there are those who would advise you to avoid response options that use certain words such as *all, none, always, never, must,* and *only,* to name a few. This, they claim, is because nothing in life is exclusive. They would advise you to choose response options that use words that allow for some exception, such as *sometimes, frequently, rarely, often, usually, seldom,* and *normally.* They would also advise you to avoid the first and last option (A and D) because test writers, they feel, are more comfortable if they put the correct answer in the middle (B and C) of the choices. Another recommendation often offered is to select the option that is either shorter or longer than the other three choices because it is more likely to be correct. Some would advise you to never change an answer since your first intuition is usually correct.

Although there may be a grain of truth in this folk wisdom, ASE test writers try to avoid this and so should you. There are just as many **A** answers as there are **B** answers, just as many **D** answers as **C** answers. As a matter of fact, ASE tries to balance the answers at about 25 percent per choice **A**, **B**, **C**, and **D**. There is no intention to use "tricky" words, such as previously outlined. Put no credence in the opposing words "sometimes" and "never," for example.

Multiple choice tests are sometimes challenging because there are often several choices that may seem possible, and it may be difficult to decide on the correct choice. The best strategy, in this case, is to determine the correct answer before looking at the options. If you see the answer you decided on, you should still examine the options to make sure that none seem more correct than yours. If you do not know or are not sure of the answer, read each option very carefully and try to eliminate those options that you know to be wrong. That way, you can often arrive at the correct choice through a process of elimination.

If you have gone through all of the test and you still do not know the answer to some of the questions, <u>then guess</u>. Yes, guess. You then have at least a 25 percent chance of being correct. If you leave the question blank, you have no chance. Your score is based on the number of questions answered correctly.

Preparing for the Exam

The main reason we have included so many sample and practice questions in this guide is, simply, to help you learn what you know and what you don't know. We recommend that you work your way through each question in this book. Before doing this, carefully look through Section 3; it contains a description and explanation of the question types you'll find on an ASE exam.

Once you understand what the questions will look like, move to the sample test. Answer one of the sample questions (Section 5) then read the explanation (Section 7) to the answer for that question. If you don't feel you understand the reasoning for the correct answer, go back and read the overview (Section 4) for the task that is related to that question. If you still don't feel you have a solid understanding of the material, identify a good source of information on the topic, such as a textbook, and do some more studying.

After you have completed all of the sample test items and reviewed your answers, move to the additional questions (Section 6). This time answer the questions as if you were taking an actual test. Do not use any reference or allow any interruptions in order to get a feel for how you will do on an actual test. Once you have answered all of the questions, grade your results using the answer key in Section 7. For every question that you gave a wrong answer to, study the explanations to the answers and/or the overview of the related task areas. Try to determine the root cause for your missing the question. The easiest thing to correct is learning the correct technical content. The hardest things to correct are behaviors that lead you to a wrong conclusion. If you knew the information but still got it wrong there is a behavior problem that will need to be corrected. An example would be reading too quickly and skipping over words that affect your reasoning. If you can identify what you did that caused you to answer the question incorrectly you can eliminate that cause and improve your score. Here are some basic guidelines to follow while preparing for the exam:

- Focus your studies on those areas in which you are weak.

- Be honest with yourself while determining if you understand something.

- Study often but in short periods of time.

- Remove yourself from all distractions while studying.

- Keep in mind the goal of studying is not just to pass the exam, the real goal is to learn!

- Prepare physically by getting a good night's rest before the test and eat meals that provide energy but do not cause discomfort.

- Arrive early to the test site to avoid long waits as test candidates check in and to allow all of the time available for your tests.

During the Test

On paper-and-pencil tests you will be placing your answers on a sheet where you will be required to color in your answer choice. Stray marks or incomplete erasures may be picked up as an answer by the electronic reader, so be sure only your answers end up on the sheet. One of the biggest problems an adult faces in test taking, it seems, is placing the answer in the correct spot on the answer sheet. Make certain that you mark your answer for, say, question 21, in the space on the answer sheet designated for the answer for question 21. A correct response in the wrong line will probably result in two questions being marked wrong, one with two answers (which could include a correct answer but will be scored wrong) and the other with no answer. Remember, the answer sheet on the written test is machine scored and can only "read" what you have colored in.

If you finish answering all of the questions on a test and have remaining time, go back and review the answers to those questions that you were not sure of. You can often catch careless errors by using the remaining time to review your answers. Carefully check your answer sheet for blank answer blocks or missing information.

At practically every test, some technicians will invariably finish ahead of time and turn their papers in long before the final call. Some technicians may be doing recertification tests and others may be taking fewer tests than you. Do not let them distract or intimidate you.

It is not wise to use less than the total amount of time that you are allotted for a test. If there are any doubts, take the time for review. Any product can usually be made better with some additional effort. A test is no exception. It is not necessary to turn in your test paper until you are told to do so.

Testing Time Length

An ASE written test session is four hours. You may attempt from one to a maximum of four tests in one session. It is recommended, however, that no more than a total of 225 questions be attempted at any test session. This will allow for just over one minute for each question.

Visitors are not permitted at any time. If you wish to leave the test room for any reason, you must first ask permission. If you finish your test early and wish to leave, you are permitted to do so only during specified dismissal periods.

You should monitor your progress and set an arbitrary limit to how much time you will need for each question. This should be based on the number of questions you are attempting. It is suggested that you wear a watch because some facilities may not have a clock visible to all areas of the room.

Computer-based tests are allotted a testing time according to the number of questions ranging from one half hour to one and one half hours. Advanced level tests are allowed two hours. This time is by appointment and you should be sure to be on time to ensure that you have all of the time allocated. If you arrive late for a CBT test appointment you will only have the amount of time remaining on your appointment.

Your Test Results!

You can gain a better perspective about tests if you know and understand how they are scored. ASE's tests are scored by American College Testing (ACT), a nonpartial, unbiased organization having no vested interest in ASE or in the automotive industry.

Each question carries the same weight as any other question. For example, if there are 50 questions, each is worth 2 percent of the total score.

The test results can tell you:

- where your knowledge equals or exceeds that needed for competent performance.

- where you might need more preparation.

Your ASE test score report is divided into content areas and will show the number of questions in each content area and how many of your answers were correct. These numbers provide information about your performance in each area of the test. However, because there may be a different number of questions in each content area of the test, a high percentage of correct answers in an area with few questions may not offset a low percentage in an area with many questions.

It should be noted that one does not "fail" an ASE test. The technician who does not pass is simply told "More Preparation Needed." Though large differences in percentages may indicate problem areas, it is important to consider how many questions were asked in each area. Since each test evaluates all phases of the work involved in a service specialty, you should be prepared in each area. A low score in one area could keep you from passing an entire test.

There is no such thing as average. You cannot determine your overall test score by adding the percentages given for each task area and dividing by the number of areas. It doesn't work that way because there generally are not the same number of questions in each task area. A task area with 20 questions, for example, counts more toward your total score than a task area with 10 questions.

Your test report should give you a good picture of your results and a better understanding of your strengths and weaknesses for each task area.

If you fail to pass the test, you may take it again at any time it is scheduled to be administered. You are the only one who will receive your test score. Test scores will not be given over the telephone by ASE nor will they be released to anyone without your written permission.

3 | Types of Questions on an ASE Exam

ASE certification tests are often thought of as being tricky. They may seem to be tricky if you do not completely understand what is being asked. The following examples will help you recognize certain types of ASE questions and avoid common errors.

Paper-and-pencil tests and computer-based test questions are identical in content and difficulty. Most initial certification tests are made up of 40 to 80 multiple choice questions. Multiple choice questions are an efficient way to test knowledge. To answer them correctly, you must think about each choice as a possibility, and then choose the one that best answers the question. To do this, read each word of the question carefully. Do not assume you know what the question is about until you have finished reading it.

About 10 percent of the questions on an actual ASE exam will use an illustration. These drawings contain the information needed to correctly answer the question. The illustration must be studied carefully before attempting to answer the question. Often, technicians look at the possible answers then try to match up the answers with the drawing. Always do the opposite; match the drawing to the answers. When the illustration is showing an electrical schematic or another system in detail, look over the system and try to figure out how the system works before you look at the question and the possible answers.

Multiple Choice Questions

The most common type of question used on ASE tests is the multiple choice question. This type of question contains three "distracters" (wrong answers) and one "key" (correct answer). When the questions are written effort is made to make the distracters plausible to draw an inexperienced technician to one of them. This type of question gives a clear indication of the technician's knowledge. Using multiple criteria including cross-sections by age, race, and other background information, ASE is able to guarantee that a question does not bias for or against any particular group. A question that shows bias toward any particular group is discarded. If you encounter a question that you are unsure of, reverse engineer it by eliminating the items that it cannot be. For example:

Under load, a diesel engine emits dark black smoke from the exhaust pipe. What should the technician check first?

A. air cleaner
B. fuel pump
C. injector pump
D. turbocharger

(A1)

Analysis:

Answer A is correct. Whenever a diesel engine emits black smoke, oxygen starvation caused by air cleaner restriction should be checked first, mainly because it can be checked and quickly eliminated as a cause. Test the intake air inlet restriction using a water manometer. The specifications should always be checked to the OEM values, but typical maximum values will be close to:

15 inches H_2O vacuum naturally aspirated engines
25 inches H_2O vacuum boosted engines

Answer B is wrong. The fuel pump (transfer pump) is very unlikely to cause a black smoke condition.

Answer C is wrong. The injector pump may cause black smoke, but it isn't the first item to check.
Answer D is wrong. The turbocharger can cause black smoke, but it isn't the first item to check.

EXCEPT Questions

Another type of question used on ASE tests has answers that are all correct except one. The correct answer for this type of question is the answer that is wrong. The word "EXCEPT" will always be in capital letters. You must identify which of the choices is the wrong answer. If you read quickly through the question, you may overlook what the question is asking and answer the question with the first correct statement. This will make your answer wrong. An example of this type of question and the analysis is as follows:

All of the following may cause premature clutch disc failure EXCEPT

A. oil contamination of the disc.
B. worn torsion springs.
C. worn U-joints.
D. a worn clutch linkage.

(A1)

Analysis:

Answer A is wrong. Oil contamination may cause slippage and disc failure.
Answer B is wrong. Worn torsion springs may cause clutch disc hub damage.
Answer C is correct. Worn U-joints will not cause premature clutch failure. This may cause driveline noise and vibration.
Answer D is wrong. Worn clutch linkage may cause disc failure due to incomplete clutch engagement or disengagement.

Technician A, Technician B Questions

The type of question that is most popularly associated with an ASE test is the "Technician A says . . . Technician B says . . . Who is correct?" type. In this type of question, you must identify the correct statement or statements. To answer this type of question correctly, you must carefully read each technician's statement and judge it on its own merit to determine if the statement is true.

Sometimes this type of question begins with a statement about some analysis or repair procedure. This is often referred to as the stem of the question and provides the setup or background information required to understand the conditions the question is based on. This is followed by two statements about the cause of the concern, proper inspection, identification, or repair choices. You are asked whether the first statement, the second statement, both statements, or neither statement is correct. Analyzing this type of question is a little easier than the other types because there are only two ideas to consider although there are still four choices for an answer.

Technician A, Technician B questions are really double true or false questions. The best way to analyze this kind of question is to consider each technician's statement separately. Ask yourself, is A true or false? Is B true or false? Then select your answer from the four choices. An important point to remember is that an ASE Technician A, Technician B question will never have Technician A and B directly disagreeing with each other. That is why you must evaluate each statement independently.

An example of this type of question and the analysis of it follows.

When discussing the adjustment of a single disc push-type clutch, Technician A says that the clutch adjustment is within the pressure plate. Technician B says that the adjustment is through the linkage only. Who is correct?

A. A only
B. B only
C. Both A and B
D. Neither A nor B   (A5)

Analysis:

Answer A is wrong. There is no provision for adjustment in the pressure plates of push-type clutches.
Answer B is correct. Only Technician B is correct. All adjustment is done through the linkage on a push-type clutch.
Answer C is wrong. Only Technician B is correct.
Answer D is wrong. Only Technician B is correct.

Most-Likely Questions

Most-Likely questions are somewhat difficult because only one choice is correct while the other three choices are nearly correct. An example of a Most-Likely-cause question is as follows:

In a twin countershaft transmission, a noise is noticeable in all gear shift positions except for high gear (direct). The Most-Likely cause of this noise is

A. a worn countershaft gear.
B. worn countershaft bearings.
C. worn rear main shaft support bearings.
D. worn front main shaft support bearings.   (B2)

Analysis:

Answers A is wrong. A worn countershaft gear would make noise whenever rotated, which is whenever the input shaft is turning.
Answer B is wrong. Worn countershaft bearings would make noise whenever rotated, which is whenever the input shaft is turning.
Answer C is wrong. A worn rear main shaft support bearing would make a noise whenever the main shaft turns, which is in every gear.
Answer D is correct. The front main shaft support bearing will only make noise when there is a speed difference between the main shaft and input shaft, which is in every gear but direct. The truck should be road tested to determine if the driver's complaint of noise is actually in the transmission. Also, technicians should try to locate and eliminate noise by means other than transmission removal or overhaul. If the noise does seem to be in the transmission, try to break it down into classifications. If possible, determine what position the gearshift lever is in when the noise occurs. If the noise is evident in only one gear position, the cause of the noise is generally traceable to the gears in operation. Jumping out of gear is usually caused by excessive end play on gears or synchronizer assemblies. This problem may also be caused by weak or broken detent springs and worn detents on the shifter rails.

LEAST-Likely Questions

Notice that in Most-Likely questions there is no capitalization. This is not so with LEAST-Likely type questions. For this type of question, look for the choice that would be the LEAST-Likely cause of the described situation. Read the entire question carefully before choosing your answer. An example is as follows:

When diagnosing an electronically controlled, automated mechanical transmission, which tool would be LEAST-Likely used?

A. a digital volt, ohmmeter
B. a laptop computer
C. a handheld scan tool
D. a test light   (B7)

Analysis:

Answers A is wrong. Digital multimeters are commonly used by technicians for diagnosing electronically controlled transmissions.
Answer B is wrong. Laptop computers are commonly used by technicians for diagnosing electronically controlled transmissions.
Answer C is wrong. A hand-held scan tool is commonly used for diagnosing electronically controlled transmissions.
Answer D is correct. A test light would be the least likely tool used for diagnosing. Test lights should not be used when diagnosing because these are not high impedance tools and may cause damage to sensitive electronic components.

Summary

There are no four-part multiple choice ASE questions having "none of the above" or "all of the above" choices. ASE does not use other types of questions, such as fill-in-the-blank, completion, true-false, word-matching, or essay. ASE does not require you to draw diagrams or sketches. If a formula or chart is required to answer a question, it is provided for you. There are no ASE questions that require you to use a pocket calculator.

4 Overview of Task List

Preventive Maintenance Inspection (PMI) (Test T8)

The following section includes the task areas and task lists for this test and a written overview of the topics covered in the test.

The task list describes the actual work you should be able to do as a technician that you will be tested on by the ASE. This is your key to the test and you should review this section carefully. We have based our sample test and additional questions upon these tasks and the overview section will also support your understanding of the task list. ASE advises that the questions on the test may not equal the number of tasks listed; the task lists tell you what ASE expects you to know how to do and be ready to be tested on.

At the end of each question in the Sample Test and Additional Test Questions Sections, a letter and number will be used as a reference back to this section for additional study. Note the following example: **A1.5.**

A. Engine Systems (12 questions)

1. Engine (2 questions)

Task A1.5 **Inspect engine mounts for looseness and deterioration.**

Example:
1. How should a truck's engine mounts be dynamically tested?
   A. Visually inspect for cracks and/or deterioration.
   B. Place the mounts under stress by jacking the engine oil pan.
   C. Observe the engine mounts while the engine is torqued with the brakes applied.
   D. Remove the mounts and perform a thorough inspection. (A1.5)

Analysis:

Question #1
Answer A is wrong. A visual inspection without applying engine torque would be inaccurate since the mounts would not be under dynamic stress.
Answer B is wrong. Jacking an engine by the oil pan could damage it, and lifting the engine will not identify defective mounts.
Answer C is correct. This is the correct way to check for worn or broken engine mounts. Placing the engine under load in first and reverse gears with the brakes applied will show movement at the worn mount as engine torque acts on the drive train.
Answer D is wrong. This procedure would be too labor intensive compared to other methods.

Task List and Overview

The decision to take a vehicle out of service can be a tough one. It is necessary to evaluate whether the mechanical problem will cause an accident or breakdown versus the need to get the load moved. Management can minimize deadline situations by listening to the driver and technician and making decisions based on the given facts and possible consequences. The driver (and technician) should perform a pre-trip inspection. The inspection should start at the left front of the vehicle and work in a counterclockwise direction around it. Permanent records are invaluable from a performance standpoint. A study of failures of individual components can alert the manager of the need to take corrective action. It also highlights components that have performed satisfactorily.

Whenever any repair procedure is performed on a vehicle, the work must be noted using a sign-off sheet before the vehicle is put back into service. If the problem noted did not require repair, but was still looked at, then this must be recorded as well. A copy of the inspection report must remain with the vehicle.

Service logs should reflect work performed on a vehicle. Active complaints/problems should always be addressed immediately with a quick visual inspection to determine if the repair can be made quickly and easily. Then check the vehicle's records if it appears that a quick repair cannot be made. All information that can be obtained could be invaluable to the timely repair of the vehicle.

The Federal Motor Carrier Safety Regulations, Part 393, states that anything mechanical that can either cause or prevent an accident is a safety item. All vehicle safety problems should be addressed, no matter how minor they appear. Section 396.19 of the Federal Motor Carrier Safety Regulations states it is the fleet manager's responsibility to ensure that all personnel doing annual inspections are qualified to perform those inspections. They should understand the inspection criteria and be able to identify defective components and systems.

The Federal Highway Administration has set up a minimum inspection standards program, under which each vehicle must carry proof that an inspection was completed. Proof can either be a copy of the inspection form kept on the vehicle or a decal. If using a decal, a copy of the inspection form must be kept on file and should indicate where an inspector can get a copy of it. The only requirement of a decal itself is that it be legible. Each vehicle must be inspected separately. This means a tractor/trailer is considered to be two vehicles, each requiring a decal; a converter dolly is a separate vehicle. The decal must show the following information: date vehicle passed inspection, name and address to contact concerning inspection record, and VIN.

Although the regulations seem to place greater emphasis on the post-trip inspection, most operators agree the pre-trip inspection is more important. Because the driver doing the inspection will be operating that particular vehicle, the incentive to ensure that the vehicle is safe is probably greater. Repairs are cheaper and less time-consuming if the driver finds them on the pre-trip before inspectors levy fines or impose penalties.

Although still a part of the overall trailer assembly, a refrigeration unit, or "reefer" as it is generally called, would not be considered an underside component of a trailer. Reefer units are commonly mounted on the upper portion of the trailer bulkhead where they can perform most efficiently.

A. Engine Systems (12 questions)

1. Engine (2 questions)

Task A1.1   Check engine operation (including unusual noises, vibration, and excessive exhaust smoke); record idle, governed rpm, and PTO rpm (if applicable).

The engine should crank when the starter is engaged at a sufficient rpm (see manufacturer's specifications). Electronic diesel and gasoline-fueled engines will not inject fuel until at least 150 rpm. Using ether can cause detonation (loud hammering sound). It should never be used on glow plug applications.

White exhaust smoke is an indication of unburned fuel exiting the exhaust pipe. Blue smoke is an indication of burning oil. Black smoke is caused by air starvation or a rich-fuel mixture. Governed rpm is the maximum engine rpm that an engine is rated for and the fuel or engine-management system will attempt to limit speeds above that setting.

Task A1.2   Inspect vibration damper.

As part of PMI, the engine is inspected, including the vibration damper. This vital engine component is mounted on the front of the engine, balancing harmonic vibrations caused by torsional loads of the rotating crankshaft. The damper is usually inspected for leaks and cracks caused by stress, dents, abrasions, and/or deterioration. A loose, damaged, or defective vibration damper can cause crankshaft failure.

Task A1.3   Inspect belts, tensioners, and pulleys; check and adjust belt tension.

All accessory belts should be checked for obvious signs of wear, tension, and/or correct application (if suspected). A belt tension gauge (as shown in the following figure) should be used whenever possible for the most accurate readings. If one is not available, then the normal deflection points at the specified locations and measurements should be observed. An idler hub should be checked for loose or worn bearings, condition of belt surfaces, cracks from fatigue or deterioration, and the correct application.

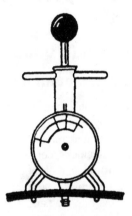

Task A1.4   Check engine for oil, coolant, air, and fuel leaks (Engine Off and Engine Running).

An engine should be checked while doing an inspection for both static and dynamic leaks. This means that the engine should be checked while it is running and while it is off. This ensures that all gasket and mating surfaces are monitored in all possible operating and nonoperating conditions. When checking for fuel leaks on an engine, it is mandatory that the engine be operating so the fuel system is pressurized, thus making a leak very apparent. A fuel leak while the engine is running is more hazardous than when the engine is off due to the higher fuel pressure with the engine running. An inspection of the cooling system should include a check of the cooling system maintenance history, coolant change intervals, component replacement(s), and past diagnoses. Although a leak in the cooling system can be a serious problem, it is not criteria for deadlining because it is not safety related.

Task A1.5   Inspect engine mounts for looseness and deterioration.

The most effective way of diagnosing a vehicle with defective engine mounts is to observe the engine mounts while torquing the engine with the brakes applied. This will allow a broken or deteriorated engine mount to be quickly observed and noted for replacement.

Task A1.6   Check engine oil level and condition; check dipstick seal.

Checking the engine oil level is a critical part of a PMI. The correct oil level is essential for proper engine lubrication and cooling. An oil level that is too low may allow air to be drawn into the pump and a loss of oil circulation, resulting in engine failure. An oil level that is too high may allow the oil to be whipped by the crankshaft, which could cause oil aeration. Aerated oil is compressible and does not provide a continuous flow of oil to the various components that depend upon oil for lubrication, cooling, or hydraulic operation. This condition could result in engine failure or accelerated wear. The recommended procedure for checking the oil level is to ensure that the vehicle is on level ground and has not been operated for at least five minutes. The dipstick must be removed and wiped with a clean rag to remove any oil from previous operation. Fully insert the clean dipstick and remove to obtain the correct reading.

Task A1.7   Check engine compartment wiring, harnesses, connectors, and seals for damage and proper routing.

Because so many systems on a modern truck depend upon the electrical circuit in one way or another, careful and frequent examination of the wiring harnesses is essential. Pay particular attention to any areas where the wiring harness contacts anything metal. Wiring harnesses should be covered with protective armor and secured in place with brackets and loom holders. Make sure all wiring is properly routed to avoid damage from moving parts and hot exhaust manifolds. Modern connectors are constructed with weatherproof seals to prevent corrosion. These seals should always be in good condition because many electrical faults are traced to problems at the connectors. When it is necessary to remove a wire from a connector for repair or replacement, be sure to use the proper tools and replacement parts. Wherever a splice or other repair is made to the wiring, it should be sealed to weatherproof it.

2.   Fuel System (2 questions)

Task A2.1   Check fuel tanks, mountings, lines, and caps.

The condition of the fuel tanks, mountings, lines, and caps are of concern during a PMI. Tanks must be checked for any signs of leakage at any point, with special attention to all seams and any damaged areas. The fuel tank mounts should be securely attached to the frame and the tank and should be free from any cracks, corrosion, and damage that could cause future failure. There should be no sign of tank shifting and all mounting bolts should be in place and tight. The tank fuel supply can also be checked for moisture contamination by the use of a rod or probe coated with water detection paste inserted into the tank. The paste is designed to change color when it comes into contact with water. If more than the tip changes color the tanks may require draining to remove the water. The fuel tank cap should be easily installed and removed and should be able to provide a good seal. Check to ensure that the cap seal is in place and in good condition. The fuel lines including the crossover line must be leak-free and should be secured by loom or brackets. Inspect all routing to ensure that no line damage will occur due to contact with other objects. If a line is chafed but no leak is present, a replacement decision should be made depending on the severity and the probability of a future leak or breakdown.

Task A2.2   Inspect throttle linkages and return springs (if applicable).

Perform throttle linkage adjustments according to manufacturer's instructions. This usually requires that the throttle lever is set back against the rear throttle stop screw before adjusting the lever position. Check the travel from idle to the full fuel position; this must meet manufacturer's specifications. If it is incorrect, adjustment of the stop screws will correct the problem. Electronic fuel-injected diesel and gasoline engines use throttle position sensors. The throttle position sensor (TPS) is a variable resistor that supplies the electronic control module (ECM) with throttle position information so the ECM can regulate engine performance and is located in the electronic foot pedal assembly (EFPA). Many systems use a 5-volt reference voltage that is sent from the ECM to the TPS, and the TPS returns a portion of it, proportional to pedal mechanical travel. The actual voltage signal returned to the ECM is converted to counts, which can be read with an electronic service tool (EST).

Task A2.3   Drain water from fuel system.

The fuel system is designed to prevent water from reaching the injection pumps and injectors. Water has poor lubricating qualities and is less compressible than diesel fuel, which means that it can cause pump damage and nozzle wear. Water separators usually have clear sumps and drain valves, which allow technicians to see and drain water from the separator when any is present. Water in the fuel tank can be drained by the use of a suction pump with a long sump that reaches to the bottom of the tank where the water would accumulate, or by a tank drain if it is so equipped. To help prevent the accumulation of water in a fuel tank, tanks should be refilled after long runs and not left near empty. The moisture that is present in the air will condense in a near-empty tank. Two terms common to diesel fuel are "cloud point" and "pour point". Cloud point refers to the temperature at which the normal paraffins in a fuel become less soluble and begin to precipitate as wax crystals. When these become large enough to make the fuel appear cloudy, this is termed cloud point. The pour point of a fuel is generally considered to be 5°F above the temperature at which the fuel will no longer flow. Pour point is generally 5°F to 25°F (3°C to 15°C) below cloud point (#2D pour point is –45°F to +5° F).

Task A2.4   Inspect water separator/fuel heater; replace fuel filter(s); prime and bleed fuel system.

The water separator can be easily serviced. When a combination water separator/primary filter is used, routine water draining and changing the filter element when the other engine filters are changed is a good practice. Some aftermarket filter elements have a longer service life and should be changed at the specified interval. Care should be taken when servicing the separator. Fuel contamination can occur easily if the canister is primed with unfiltered fuel or dirt is allowed to enter when the element is being changed. Most fuel filters are changed at routine intervals (PMI schedules). These filters must be primed with fuel before the engine can be started. Most fuel filters can be filled with clean fuel and reinstalled, but some manufacturers recommend that the secondary filter be installed dry and primed with a hand pump. This procedure ensures that only filtered diesel fuel enters the secondary filter. When hand priming is necessary, a line fitting upstream from the pump should be cracked open and priming should continue until no air bubbles are present at the fitting. The fitting should be closed and the engine should be cranked for short periods of time with adequate breaks to allow for starter motor cooling. Once the engine starts, it should be operated until it runs smooth with no stumbling. At this point a nonintegral hand pump should be removed.

Task A2.5   Inspect crankcase ventilation system.

The PCV valve should always be checked when any inspection of the emission control system is performed. Improper operation of the PCV can cause erratic engine operation as well as higher than normal gasoline exhaust emissions, thereby possibly causing fines and/or penalties due to violations.

Task A2.6   Inspect gasoline and diesel emission control systems, including exhaust gas recirculation (EGR), secondary air injection (AIR), air inlet temperature control, early fuel evaporation (EFE), evaporative emission (EVAP) systems, and catalytic converter.

One of the earliest methods used to reduce the amount of hydrocarbons (HC) and carbon monoxide (CO) in the exhaust of gasoline engines was by forcing fresh air into the exhaust system after combustion. This additional fresh air causes further oxidation and combustion of the unburned hydrocarbons and carbon monoxide. The system can be equipped with or without an external air pump. Without it, the system can be referred to as a pulse, aspirator, suction, or reed air injection.

EGR is a method of reducing oxides of nitrogen (NO_x) by circulating a small amount of exhaust gases back through the intake manifold to dilute the intake charge and lower the combustion temperature. The EGR system should only function on a warm engine, operating above idle. A common test of a vacuum-operated EGR valve is to apply vacuum to the valves diaphragm at idle, which should produce a rough idle or even stall out. At above-idle operation, vacuum should be present at the valve. Air inlet temperature control is simply a dual position door located in the air cleaner inlet, which allows direct inlet air in when the engine is warm and preheated air in when the engine is cold. The preheated air is air drawn past the exhaust manifolds through ducting and shrouds. EFE is a process of diverting exhaust gases through a passage in the intake manifold during cold engine operation, to prevent fuel condensation on the cold walls. Some EFE systems use an electric heating grid to warm the intake. EVAP use a sealed fuel tank with pressure relief vents that are plumbed to a carbon canister. The carbon canister absorbs the carbon from the gasoline vapor. When the engine is running, manifold vacuum draws the gasoline vapor into the engine to be burned. Catalytic converters are used on most gasoline engines. Catalytic converters use catalyst substances to convert harmful emissions into nonharmful emissions. Palladium and platinum are oxidation catalysts that attempt to oxidize hydrocarbons and carbon monoxide into water and carbon dioxide. Rhodium is a reduction catalyst that tries to reduce oxides of nitrogen into the elements nitrogen and oxygen.

Generally diesel engines have used less external emission control devices. Computer-controlled diesels have very precise control over the combustion process, so most engines meet emission standards. An increasing number of diesel engines are being equipped with cooled-exhaust gas recirculation (C-EGR) to meet more stringent emission standards. Single stage or oxidizing catalytic converters are used on some engines while others use particulate traps to filter soot particles. To reduce cold engine emissions, some engines use intake air heaters to reduce unburned hydrocarbons. Catalytic converter operation can be checked with an infrared thermometer. A correctly operating converter should operate in the range of 1500 to 1800°F and the inlet temperature should be less than the output temperature. All of the above systems can be checked and serviced relatively easily, but with such a broad range of systems, manufacturers' manuals should be consulted for correct service procedures.

3. Air Induction and Exhaust System (2 questions)

Task A3.1   Check exhaust system mountings for looseness and damage.

To check exhaust pipes and support brackets, a pry bar can be used to check for looseness. All mounting brackets should be examined for cracks and clamps must be tight. The complete system should be visually inspected for rust, damage, and cracks.

Task A3.2   Check engine exhaust system for leaks, excessive noise, proper routing, and missing or damaged components (heat shields and guards).

Any exhaust system that is leaking in front of or below the driver/sleeper compartment and the floor pan has potential for cab entry of exhaust fumes. These fumes could result in a drowsy or asphyxiated driver, which could cause an accident. The exhaust routing must be checked and assembled in a way that does not allow contact of the exhaust system with any wiring, air lines, fuel lines, or combustible materials. The CVSA out-of-service criteria would deadline a vehicle for any of these faults.

Task A3.3   Check air induction system routing, piping, charge air cooler, hoses, clamps, mountings, and indicators; check for air restrictions and leaks.

Depending on the conditions the vehicle is driven in, truck air induction ducts do not generally have to be cleaned out on every PMI. It is always a good idea to give a full visual inspection to the air induction system for blockage and/or deterioration whenever a PMI is performed, only cleaning it when necessary. The air inlet restriction should be checked with a water manometer or accurate vacuum gauge. If the restriction exceeds the maximum specification, the air filter should be replaced. Restricted dry filters should not be tapped on the floor or blown with air pressure to dislodge dirt. Both of these practices can result in filter damage and dirt bypassing the filter if it is reinstalled.

Task A3.4   Inspect turbocharger for noise and leaks; check mountings and connections, check wastegate, linkages, and hose.

A turbocharger operates on the principle of thermal expansion to drive a turbine wheel, located in the turbine housing. This in turn spins the compressor wheel in the impeller-housing. The compressor wheel is located in the inlet side and draws in fresh air and pressurizes it. It is necessary to contain this pressurized air in hoses, manifolds, and an intercooler. A turbocharger can operate within manufacturer's specifications and still have acceptable leakage. To test for proper leak rate, follow the most current information provided in manufacturers' published service bulletins and do not exceed the recommended test air pressure. The clamps used are spring loaded and should be torqued to specifications and placed in a way as to not chafe through other hoses or components with engine movement. Modern diesel engine turbochargers perform at very high speeds and can reach well over 100,000 rpm. When running the engine without the air-intake, be sure to install an intake guard over the turbocharger inlet.

Task A3.5   Service or replace air filter(s) as needed.

Engines require approximately 9,000 to 10,000 gallons of air by volume for every gallon of fuel consumed. This results in a volumetric air-fuel ratio of approximately 10,000 to 1 for both gasoline and diesel engines. Air-fuel ratio by mass is approximately 15:1 and is referred to as the stoichiometric ratio. With this large volume of air entering an engine it is important to provide good filtration, as airborne dust, sand, and other debris ingested into the cylinders will lead to premature engine failure (called "dusting" caused by loose clamps or ill-fitting air filter assemblies). Obstructions such as a restricted air filter will reduce airflow/volume and cause an engine to operate in a rich (fuel) condition and cause cylinder wash down of the oil film on the cylinder walls, again leading to engine failure. Always service following manufacturer's specifications and intervals. Typical restriction specifications are 15 inches of water vacuum for naturally aspirated engines and 25 inches of water vacuum for boosted engines.

4.   Cooling System (4 questions)

Task A4.1   Check operation of fan clutch (viscous/thermostatic, air, and electric).

Fan clutch types may be viscous, thermatic, air apply, air release, electric, and hydraulic. Each type has different characteristics and should first be identified to avoid improper diagnosis. Viscous and hydraulic units should not exhibit any signs of fluid leakage. Most pneumatic fan clutch failures are caused by air leaks. Whatever PM level is being performed, an air leak check is critical to maintaining proper operation of the pneumatic fan clutch. The engine computer may control on-off operation. Fan clutches may be engaged by engine compartment temperatures, fluid temperatures, or by engine management strategies.

Task A4.2   Inspect radiator (including air flow restriction, leaks, and damage) and mountings.

Radiators, because of their location mostly at the front of the vehicle, are susceptible to damage from airborne objects and are thus prone to damage or restriction (i.e., bent cooling fins, corrosion, or in the event of large objects, punctures causing loss of coolant). It is important to inspect the cushion mounts at service intervals to prevent vibration and other movement damage of the radiator and related components. The radiator is usually mounted close to the engine with intercooler, A/C condenser, and oil coolers inhibiting airflow to it. Separate these units at service intervals to inspect for a build-up of compacted road debris that can severely compromise the radiator's ability to effectively cool and rinse out with low-pressure water.

Task A4.3   Inspect fan assembly and shroud.

The fan is subject to receiving road debris created by the ram air effect of road speed and the vacuum of the fan rotation. This suspended particulate travels through the radiator and impacts the fan blades causing a sandblast effect on the fan blades, clutch, and related components assemblies. Riveted fan blades can loosen. Plastic fan blades can become weakened and crack. Fan clutches can become inoperative. Fan blades can dislodge while performing inspections; check blade condition before running engine. Inspect for weak or broken engine/transmission mounts, which can lead to fan blade and shroud damage. Check for proper fan blade clearance, alignment, engagement, and engine/transmission mount function while performing preliminary tests. Also shroud clearance and flow restriction may affect fan efficiency.

Task A4.4   Pressure test cooling system and radiator cap.

The normal system pressure for a cooling system is in the range of 15 psi, depending on the particular design of the vehicle being tested. The purpose of the radiator cap is to allow a higher operating temperature of the engine by raising the boiling point of the coolant (3°F for each 1 psi). Therefore, it is important to test the radiator cap to ensure that it will retain the indicated pressure setting. Using a pressure tester unit, install the cap onto the tester and proceed to apply pressure to the specified pressure release point. If the specified limit cannot be achieved then cap replacement is required. An underperforming cap may cause engine overheating and coolant loss. If the coolant system has a recovery tank then the cap should have two seals. The one on the end of the coil spring is to contain fluid and the rubber seal under the cap lid is to create a vacuum and aid in returning coolant from the overflow container to the engine.

Pressure testing the engine cooling system for leaks is performed in the same manner. Apply pressure by means of the same tester unit. The gauge reading should achieve the same setting as the recommended radiator cap. The gauge reading should hold steady. The gauge should not increase other than a flutter (water pump pressure surge) upon startup. A large increase usually indicates compression gas infiltration into the cooling system, or if the coolant is oil discolored then check for oil entering the coolant.

Task A4.5   Inspect coolant hoses and clamps for leaks, damage, and proper routing.

All heater and water hoses should be sound and pliable. All clamps are checked for tightness and/or deterioration. A visual inspection of the entire cooling system is always a good practice.

When inspecting hoses, remember that hoses deteriorate on the inside before any sign of wear occurs on the outside. Hoses should be checked by looking inside the hose for cracks and splits. Silicone hoses are easily damaged by overtightening the clamps. They require special clamps that should be torqued carefully. When correctly installed they provide better sealing and longer service life. Some coolant leaks occur only when the system is cold. They are sealed by expansion during operating temperatures. Check for leaks both when cold and at operating temperature.

Task A4.6   Inspect coolant recovery system.

The radiator cap will maintain the pressure within the cooling system at or below approximately 15 psi. During heavy loads and prolonged periods of idling, this pressure may exceed the rating of the radiator cap and fluid will be vented to an expansion tank or bottle. Once enough fluid has been vented to reduce the pressure within limits, the spring tension of the radiator cap will close the vent line. At normal temperatures with the pressure cap removed small fluctuations in coolant level can be expected. Inspection of this fluid level may indicate deeper problems. A constant overflowing of the fluid from the tube may indicate air pockets, uneven heating, or hot spots in the engine. Repetitious and rhythmic surges in this level usually indicate a leak to the combustion cylinder, such as a blown head gasket.

Task A4.7   Identify coolant type. Check coolant for contamination, supplemental coolant additives (SCAs), and protection level (freeze point).

Coolant freeze and boil protection levels can be tested with a hydrometer or refractometer. Different antifreeze to water ratios are required to raise or lower the protection level depending upon the temperatures the vehicle will encounter. A 50/50 mixture is most common. To be compatible with the different gaskets and metals that are in various engines, current trucks use one of three types of antifreeze that may be identified by a label in the engine compartment: ethylene glycol, propylene glycol, or carboxylate, which is an extended-life antifreeze. Always ensure that the proper tester is used for the type of antifreeze used.

Generally OEMs recommend that the coolant SCA level be tested at each oil change interval. Additionally, whenever there is a substantial loss of coolant and the system has to be replenished, the SCA level should be tested. Each test provided by an OEM is designed to monitor the SCA package required for its product and cannot generally be used for other OEM's product. Also, the test kits usually consist of test strips that must be stored in airtight containers that have expiration dates that should be observed. Coolant test kits permit the technician to test for the appropriate SCA concentration, the pH level, and the total dissolved solids (TDS). The pH level determines the relative acidity or alkalinity of the coolant. Acids may form in engine coolant exposed to combustion gases or in some cases when cooling system metals (ferrous and copper bases) degrade. The pH test is a litmus test in which a test strip is first inserted into a sample of the coolant, then removed and the color of the test strip indexed to a color chart provided with the kit. The optimum pH window is defined by each OEM, but normally falls between 7.5 and 11.0 on the pH scale. Higher acidity (readings below 7.5 on the pH scale) readings in tested coolant are indications of corrosion of ferrous and copper metals, coolant exposure to combustion gases and in some cases, coolant degradation. Higher than normal alkalinity readings indicate aluminum corrosion and possibly that a low silicate antifreeze is being used where a high silicate antifreeze is required.

Task A4.8   Service coolant filter/conditioner.

Coolant filters have service intervals and manufacturers' literature should be consulted to determine the proper maintenance schedule. This is important on diesel engines as filters become restricted and coolant additives break down due to extended operation. Not complying with this service information will result in premature engine failure. Some coolant filters contain an SCA charge and corrosion inhibitors. These filters should only be replaced as directed. Never over-service coolant filters that contain an SCA charge.

Supplemental coolant additives are a critical ingredient of the coolant mixture. The actual SCA package recommended by an engine manufacturer depends on whether wet or dry cylinder liners are used, the materials used in the cooling system components, and the fluid dynamics (high flow/low flow) within the cooling system. The operator may wish to adjust the SCA package to suit a specific operating environment or set of conditions. Abnormally hard water, for instance, will require a greater antiscale protection. Depending on the manufacturer, SCA may be added to a cooling system in a number of ways. The practice of installing the SCA in the system coolant filter is less commonly used today as it resulted in generally higher levels of additives than required and an excess of additives can create problems. Most OEMs suggest testing the SCA levels in the coolant followed by adding SCA to adjust to the required values. Never dump unmeasured quantities of SCA into the cooling system at each PMI.

Task A4.9   Drain and refill cooling system; bleed air from system; recover coolant and dispose of in accordance with federal, state, and local regulations.

Cooling system maintenance requires periodic draining and refilling of the system. When doing this it is important to know that all coolants are alike and are not always compatible with each other. Identify the coolant being used and be sure to follow the pertinent change intervals and refill with the correct type. ELCs have a longer service life than EG and PG coolants. When refilling the system it is necessary to bleed the system to allow trapped air to escape. This can be done by opening a hose or fitting on the top of the engine before the thermostat and filling the system until liquid coolant appears. This prevents trapped air from creating a hot spot when the engine is first started. Coolant that has been drained should be handled and stored in accordance with federal, state, and local regulations. This could include on-site recycling or removal by a licensed service. There should be an MSDS sheet posted in your work area that gives information regarding proper handling of each coolant used.

Task A4.10   Inspect water pump for leaks and bearing play.

The water pump is usually mounted on the front of the engine block with a gasket between the block and the water pump. A leaking water pump seal causes coolant to drip from the drain hole on the bottom side of the pump. Coolant leaking past the water pump seal can lead to bearing failure. A defective water pump bearing may cause a growling noise with the engine idling.

5. Lubrication System (2 questions)

Task A5.1   **Change engine oil and filters; visually check oil for coolant or fuel contamination; inspect and clean magnetic drain plugs. Dispose of used oil and/or filters in accordance with federal, state, and local regulations.**

After the initial lube change at 1,000 to 3,000 miles, subsequent lube changes should be made at the engine manufacturer's suggested intervals for the oil type used, dependent on Class A or AA highway operation. Class A operation is defined as driving on well-maintained highways of concrete or asphalt construction.

Oil filters trap particles that would otherwise cause damage to critical components like engine bearings. Regular replacement of filter elements and inspection of the housing and other component parts is necessary. Modern engines use a variety of materials from cast iron to diecast aluminum for the housing and steel or plastic for machined parts. For this reason, the technician is limited to a visual inspection of filter assembly parts. The visual inspection must be complete and thorough. If there is any doubt, or a component is questionable, replace it. Cracks anywhere in the housing and nicks on machined surfaces can cause major engine failures and, as such, must be detected and appropriate action taken to correct the problem. Some filter assemblies incorporate a bypass valve in their housing. These valve assemblies must be removed and inspected for corrosion, wear, and other signs of damage during overhaul or any time the filter assembly contains metal particles. Canister filters should be cut open and the element material examined for content. Suspended metal particles indicate a potential for bearing failure.

A milky gray sludge indicates water in the oil caused by a possible head or cylinder block leak, which may already have caused bearing damage. Ensure all gasket surfaces are straight and free of nicks, which could cause an improper seal of the assembly. On element-type filter assemblies, ensure that housing grooves, center bolt threads, springs, metal gaskets, and spacers are in good condition. Inspect for distortions in the canister, cracks around the canister bolt hole and clogged passageways in the bolt. Reinstall the bypass valve according to manufacturer's installation procedures and set for proper pressure. Replace the element and all gaskets and supplied seals. When reassembling filter assemblies, apply a small amount of oil to all seals to ensure a tight seal. Prefilling of canister-type filters is recommended by most OEM's to ensure adequate oil is present during startup to lubricate the bearings.

Used oil and filters are considered hazardous waste and must be disposed of in accordance with federal, state, and local regulations. Used oil filters should never be placed in with other trash but should be disposed of properly. Filter crushing equipment is available to reduce the volume of disposed of filters. In some states used oil can be burned in a waste oil furnace, but the appropriate method of disposing is removal by licensed recycling services. Used oil can be harmful to your skin and you should take steps to minimize contact with it. Consult the MSDS sheet posted in your work area concerning the handling of motor oils.

Task A5.2   **Take an engine oil sample.**

During a normal PMI, an engine oil sample is taken for inspection of contamination, metal, and dirt content. Also, corrosion-causing agents such as sulfuric and hydrochloric acids and some esters are sometimes tested for content as well. It is for this reason, that the engine oil sample be taken at normal operating temperature to ensure that all possible contaminants and corrosives are measured in the oil in its normal operating condition.

The oil sample should be taken during an oil change at the midpoint of the drainage. The test bottle should be clean before the sample, and sealed and labeled afterward. The label should contain all the pertinent data about the engine and the oil used. Low-level testing can be performed in the shop with a blotter test kit. This kit will compare the darkness and fluidity of a drop of oil on the blotter paper to a code chart that identifies the level of dirt, soot, and particles found in the oil, and the color change of the oil will indicate the acidity. A more comprehensive test can be performed at a specialized lab or oil manufacturer's laboratory. These labs perform spectrographic tests, which identify and quantify the contaminants found in the oil.

B. Cab and Hood (6 questions)

1. Instruments and Controls (2 questions)

Task B1.1   Inspect key condition and check operation of ignition switch.

Inspect the vehicle keys for wear, nicks, bends, or other damage. Ensure that the ignition switch is securely mounted and operates smoothly without binding and provides constant power at the accessory, on, and crank positions. If the truck is equipped with a push button start switch, it should operate easily without sticking while properly closing the cranking circuit when depressed. Some trucks with keyless ignition systems require a driver ID code or card instead of a key.

Task B1.2   Check operation of indicator lights, warning lights and/or alarms.

Warning lights have changed significantly over the years. Earlier lamps were small, round lights that were color coded for their system; yellow for a warning or caution in a non-safety-related system and red for safety-related system warnings. Major or safety-related systems may still use lights but many non-safety systems use LEDs behind a pictogram of the system it is monitoring. This improvement allows for less electrical current usage in the instrument panel and more reliable operation. Most vehicles have a self-check feature that illuminates all warning lamps when the ignition switch is first switched to the on position. This lets the driver and/or technician know that the lamp and circuit is operational. Alarms or buzzers are also used for some systems such as low air pressure (which is mandatory) and for low oil pressure. The low air pressure alarm must sound when the system air pressure drops to below 50 percent of the governor cut-out pressure (approximately 60 psi).

Task B1.3   Check operation of instruments/gauges and panel lighting.

A short circuit in the sending unit wire to the gauge will lower resistance in that circuit, causing the gauge needle to read high. Generally a short to ground in a sending unit wire will cause an overload of current to the gauge, which will consequently burn out in a very short time. A typical modern oil pressure gauge is more operator friendly—using symbols instead of words to communicate a vehicle's status. For example, an oil can (instead of the words "oil pressure") may be used to convey a status report on the vehicle's oil pressure.

Although not considered official criterion for out-of-service (OOS) status, a malfunctioning oil pressure gauge may be cause for an inspector to deadline a vehicle because a no-oil pressure condition can rapidly destroy an engine. Also, a proper and thorough check of a vehicle's instrument panel is always an important part of any PMI. When the ignition is first turned on, electronic gauges perform a self-test by sweeping the needles.

Task B1.4   Check operation of electronic power take off (PTO) and engine idle speed controls (if applicable).

Modern diesel engines manage idle speed and shutdown by computer control. This is sometimes a PTO function when the engine speed is set to operate auxiliary equipment. Idle speed usually can be raised and lowered by the use of a toggle switch provided on the dash. During an inspection, check the operation of this switch to ensue that idle speed can be controlled. If equipped with a PTO switch, also check its operation.

Engine shutdown can be programmed to occur after a desired time interval. When the desired time has elapsed without input from the throttle pedal or the clutch pedal the engine will shut down automatically. To change this programmed time, an EST is needed. It is important to minimize idle time to prevent wasting fuel and to prevent costly wear to the engine. Some drivers like to idle the engine for extended periods to operate the heater or air conditioner in the sleeper, but this is not a good practice. Most engine management systems have automatic stop/restart features that expand the normal idle shutdown feature to further reduce idle time. They will automatically stop and restart the engine when required to hold engine temperature above a programmed specific value, maintain the battery charge level, and to keep the cab/sleeper at a temperature level programmed into customer parameters. Other benefits claimed are reduced emissions, noise, and engine life.

In order for the stop-restart feature to be enabled, a number of conditions must be met. One OEM system requires that the ignition switch be on, the engine idling with the idle shutdown timer enabled, the hood closed, the transmission in neutral or park, the parking brake set, and the master cruise switch in the on position. When the system determines to start the engine, based on sensor inputs, the active light illuminates and an alarm in the engine compartment sounds and the engine cranks. In the event the engine fails to start, a second attempt will occur in 45 seconds. At startup, engine rpm ramps up to 1000 rpms. If the restart was prompted by input from the cab temperature sensors, after about 30 seconds, the climate control system will turn on to manage cab/sleeper temperature.

Engine shutdown can also occur to prevent damage to the engine from lack of lubrication or overheating conditions. The way in which the engine electronics respond to a potentially damaging condition is known as failure strategy. Failure strategy response levels are customer programmable and typically would be driver-alert (lowest level), engine ramp-down, and engine shutdown (highest level). Manufacturers recommend programming the highest level failure strategy response but, for obvious reasons, emergency vehicles such as fire trucks are usually programmed with low-level failure strategy. An EST can be used to check the programmed failure response.

Task B1.5   Check operation of defroster, heater, ventilation, and A/C (HVAC) controls.

Check the HVAC system to see if it is able to regulate temperature in the cab and sleeper. The blower motor should operate in all speeds. The blower should produce an adequate flow of air and should not be abnormally noisy. Move the mode selector, making sure that air is directed to the selected outlets. A pocket thermometer can be used to check the outlet temperature of the heater and the defroster. With the engine idling, you should hear the A/C compressor clutch engage when air conditioning is selected. Raise the engine to about 1500 rpm and close the doors and windows. Turn the blower to high and select the coldest temperature. Run the system for about 10 to 15 minutes and compare the temperature of the air at the outlet nearest the evaporator with specifications in the service manual. Check the operation of the sleeper controls in a similar manner. If the truck is equipped with automatic temperature control (ATC), the outlet temperature should be within several degrees of the set temperature.

Task B1.6   Check operation of all accessories.

Identify all the cab and chassis accessories. Where possible, switch on to verify operation. Many accessories can be checked during a road test. For instance, a VORAD (vehicle on-board radar) CWS (collision warning system) has to be tested on the road as well as with an EST.

Task B1.7   Using diagnostic tool or on-board diagnostic system, extract engine, transmission and brake monitoring information and codes.

Today's diesel trucks contain a variety of computerized control systems to operate engines, transmissions, antilock brakes, etc. According to Society of Automotive Engineers (SAE) Standard J1930, diagnostic codes are known as diagnostic trouble codes or DTCs. DTC access may be by blink codes or read off the chassis databus. You need to understand how to check, record, and clear

DTCs. The typical process is to use a handheld EST or laptop computer to access the databus; then check for, record, and clear DTCs. These systems are controlled and monitored by electronic modules. Most systems will accommodate a snapshot readout from the ECM to facilitate troubleshooting intermittent problems that either do not generate codes or do not store a code without a clear reason. Control modules store fault codes against parameter identifications (PIDs). To access the databus, the technician connects a diagnostic reader to the ATA connector. Once the diagnostic reader is connected, most systems allow two-way communications for reading system parameters, calibration data, diagnostics, data programming, recalling faults, and audit trails. Some electronic systems can display active fault codes by blinking two-digit numerals on the instrument panel indicator. Diagnostic tests are menu-driven and should be used to identify the FMIs (Fault Mode Indicators). To test a system, once the problems have been corrected, the now inactive fault codes need to be cleared. ESTs can be used to clear these codes.

ESTs capable of reading the chassis databus are connected to the on-board electronics by means of an SAE/ATA (American Trucking Association) J1587/J1708/1939 6- or 9-pin Deutch connector in all current systems. This common connector and the adherence by the engine electronics OEMs to SAE databus software protocols enables proprietary software of one manufacturer to at least read the parameters and conditions of their competitors. This means that if a Catapiller-powered truck has an electronic failure in a location where the only service dealer is DDC, some basic problem diagnosis can be undertaken using the DDC electronic diagnostic equipment.

There are five categories of ESTs as follows:

1. On-board diagnostic lights
   Blink or flash codes (OEMs use both terms) are an on-board means of troubleshooting using a dash or ECM-mounted electronic malfunction light or CEL (check engine light). Normally only active codes (ones indicating a malfunction at the time of reading) can be read in truck (engine) ECMs but some will also read historic codes.

2. Scanners
   These are read-only tools are capable of reading active and historic or inactive (logged but not currently indicating a malfunction) codes and sometimes system parameters, but little else. They are obsolete as a truck/bus diesel engine diagnostic tool.

3. Generic reader/programmers
   A microcomputer-based EST designed to read and reprogram all proprietary systems in conjunction with the appropriate software cartridge or optical card. They are usually tough and portable. The ProLink 9000 has become the industry standard.

4. Proprietary ESTs
   These are usually PC-based test instruments packaged by the OEM for use exclusively on their system. They have the advantage of offering an optimum degree of user friendliness and the disadvantage of being system specific and high in cost. Examples are:

   • Cummins: Compulink, Echeck

   • Cat: ECAP

   • DDC: DDR Programming Station

Note: These propietary ESTs have fallen out of favor due to the near universal acceptance of Item #5.

5. Personal computers
   Most OEMs have made the generic PC and its operating systems (MS Windows) their diagnostic and programming tool of choice.

PCs are easily upgradable and have vast computing power when compared with proprietary reader/programmers. OEMs can upgrade/update software more easily and the PC is not confined to the limitations of all the other existing electronic troubleshooting tools. PCs are connected to vehicle ECM through the SAE/ATA J1708/1939 6- or 9-pin connectors; a serial link or interface module may also be required, depending on the age of the system.

Truck engine OEMs have chosen to use MS Windows-driven programs and offer comprehensive courses on their own management systems, which usually include a thorough orientation of PCs.

Generally, SAE J1587 covers common software databus protocols in electronic systems, SAE J1708 covers common hardware protocols, and the more recent SAE J1939 covers both hardware and software protocols databus. The acceptance and widespread usage of these protocols enables the interfacing of electronic systems manufactured by different OEMs on truck and bus chassis, plus permits any manufacturer's software to at least read other OEMs' electronic systems.

2. Safety Equipment (1 question)

Task B2.1   Check operation of electric and air horns.

Electric and air horns are important safety devices enabling the operator to warn other motorists of imminent danger. Both are required to be fully functional. They should be tested during the driver's pre-trip and post-trip inspections. Verify their operations as part of each preventive maintenance inspection.

Task B2.2   Check condition of safety equipment, including flares, spare fuses, reflective triangles, fire extinguisher and all required decals.

Check that the truck is equipped with all the necessary safety equipment. The truck should have spare fuses (unless equipped with circuit breakers), three red reflective triangles, and a properly charged and rated fire extinguisher within arm's reach of the driver's seat. Flares, lanterns, and flags are optional.

CVSA out-of-service criteria calls for deadlining when safety devices such as chains and hooks have been repaired improperly. Federal regulations prohibit the usage of a vehicle that does not have at least one light illuminated for loads projecting more than 4 feet beyond the vehicle.

Rear chains and hooks that are used in double trailer systems for hookup should not be worn to the extent of a measurable reduction in link cross sectional area or have any significant abrasions, cracks, or other faults that would affect structural integrity. If any defects are found, deadline the vehicle.

Task B2.3   Inspect seat belts and sleeper restraints; check supplemental restraint system (SRS) warning light operation (if applicable).

Check the operation of all seat belts and restraints. With the seat belt on and the truck moving, activate the parking brake to brake the truck very quickly and determine if the seat belt inertia reel locks. Visually inspect seat belt anchors, seat bolt downs, and sleeper bunk harnesses.

If the vehicle is equipped with the supplemental restraint system (SRS), you should also check the operation of that. This is done easily because whenever the key is turned on, the SRS performs a self-test. You should observe that the "SRS" or "Airbag" indicator light illuminates and stays on for several seconds following startup—then extinguishes. If it stays on continuously or blinks, there is a problem in the system that needs correction.

Task B2.4   Inspect wiper blades and arms.

Check the wiper blades and arms for correct operation. The blade should remove water from the entire sweep area. Replace the blade if the rubber is cracked, separated, has segments missing, or is stiffened with age and cannot properly clear the windshield. The arm attaches securely to the wiper mechanism and provides the correct tension for the wiper blade to windshield contact. The arm should be replaced if it cannot perform as stated above or if it is bent. CVSA safety standards out-of-service criteria state that any wiper unit that has inoperable parts or missing wipers that render the system ineffective on the driver's side shall be cause for deadline.

Task B2.5   Check wiper and washer operation.

Check the wiper and washer systems for proper operation. Ensure that the wiper blades and arms are in good condition and that the linkage is lubricated and in good working order. Check electric wiper motor operation through all the speeds and delay settings. If any speeds are inoperable, follow the manufacturer's service procedures to identify the failed component and its replacement procedures. Air-operated wiper motors should be checked as previously mentioned but with additional checks for air leaks. Adjust the arm to sweep and park in the correct areas and not contact the cab in any position. Inspect the washer fluid container for debris that could impair the future operation of the system and clean if necessary. Ensure that the correct washer fluid is in use for the season and conditions that the vehicle is operating under. Check the condition and aim of the spray nozzles and perform any necessary adjustments or required replacement of defective components. CVSA safety standards out-of-service criteria state that any wiper unit that has inoperable parts or missing wipers that render the system ineffective on the driver's side shall be cause for deadline.

Task B2.6 Check for all required vehicle permits, registration, decals, and inspection papers.

Every vehicle must have a registration (license) in order to operate. The majority of trucks or truck-trailer combinations weighing more than 26,000 pounds will be registered under the International Registration Plan (IRP). A cab card is issued to the vehicle, stating the IRP areas in which the vehicle is allowed to operate.

The Federal Highway Administration has set up a minimum inspection standards program, under which each vehicle must carry proof that an inspection was completed. Proof can either be a copy of the inspection form kept on the vehicle or a decal. If using a decal, a copy of the inspection form must be kept on file and must indicate where an inspector can get a copy of it. The only requirement of a decal itself is that it be legible. Each vehicle must be inspected separately. This means a tractor/trailer is considered to be two vehicles, each requiring a decal; a converter dolly is a separate vehicle. The decal must show the following information: date vehicle passed inspection, name and address to contact concerning inspection records, and the VIN.

3. Hardware (1 question)

Task B3.1 Inspect windshield glass for cracks, chips or discoloration; check sun visor operation.

Check the windshield glass for cracks, dirt, illegal stickers, or discoloration that obstruct the driver's view. Clean and adjust as necessary. Any crack over $\frac{1}{4}$-inch wide, intersecting cracks, discoloration, or other vision-distorting matter in the sweep of the wiper on the driver's side will take a vehicle out of service by the commercial CVSA out-of-service standards.

Task B3.2 Check seat condition, operation, mounting, and suspension components.

Many cab air seats contain one or two air springs, two shock absorbers, and a leveling valve. The leveling valve maintains the proper air pressure in the air springs to provide the correct seat height. An air-suspended seat reduces road shock transferred through the chassis, cab, and seat to the driver. Air pressure is supplied to the air-suspended seat through a pressure protection valve and a regulator. You adjust the height control, or leveling valve (similar to air suspension systems), to maintain the proper seat height.

Task B3.3 Check door glass and window operation.

Check to see if the windows will go up and down. Windows must go up and down when actuated electrically or by hand.

Task B3.4 Inspect steps and grab handles.

Inspect that all steps are in good condition and secure. Corrugated steps must be level and should support 300 pounds. Check that the grab handles are also secure and tight to the cab.

Task B3.5   **Inspect mirrors, including mountings, brackets, glass, heaters, and motors.**

A typical fault such as a loose mirror mounting is always recorded on a PMI. There should be no blemishes on the mirror glass. This is in accordance with the Department of Transportation's Federal Motor Carrier Safety Regulations and the CVSA standard inspection procedures.

If the truck is equipped with either heated or motorized mirrors, you should also make sure they are functioning. When the mirror heat switch is on, an indicator light should be on and the mirror surface should become warmer than room temperature. When equipped with motorized mirrors, toggle switches allow you to adjust the position of either mirror from the driver seat.

Task B3.6   **Inspect and record all observed physical damage.**

Check the vehicle area for hazards on the vehicle and any physical damage such as dents or cracks. Performing a driver's circle check is a good way of doing this. Use a combination of visual and hands-on inspection. The inspection should start at the left front of the vehicle and work in a counterclockwise direction around it.

Task B3.7   **Lubricate all cab and hood grease fittings.**

Use light machine oil and a lithium-based grease on hinges and latches.

Task B3.8   **Inspect and lubricate door and hood hinges, latches, strikers, lock cylinders, linkages, and cables.**

Lubrication of door latches, locks, and hinges are always performed as part of a PMI. Light machine oils and lithium base grease can be used.

Task B3.9   **Inspect cab mountings, hinges, latches, linkages; service as needed.**

Most current cabs use air ride. Check the foreword pivots for wear and excessive movement, check the ride height and leveling valve, inspect shock absorbers for leakage. Visually and dynamically test the bushings.

Task B3.10   **Inspect tilt cab hydraulic pump, lines, and cylinders for leakage; inspect safety devices; service as needed.**

With most hydraulic cab lift systems there are two circuits: the push circuit that raises the cab from the lowered position to the desired tilt position and the pull circuit that brings the cab from a fully tilted position up and over the center.

Inspect the hydraulic cylinders for external leakage and determine if the cab can be raised and lowered. Check for internal leakage by observing bleed down. Make sure the mechanical stops engaged and hold safely.

Task B3.11   **Check accelerator, clutch, and brake pedal operation and condition.**

Check the accelerator linkage for binding and lubricate it at the appropriate locations. Test the brake pedal for operation and lubrication. Check clutch-pedal free-play and if an adjustment is performed the work must be noted using a signoff or work performed sheet before the vehicle can be put back into service. If the problem noted did not require repair, but was still looked at, then this must be noted as well. A common problem in electronic accelerator assemblies is dirt build-up around the TPS assembly.

Task B3.12 Check cab ride height. Inspect cab air suspension springs, mounts, hoses, valves, shock absorbers, and fittings for leaks and damage.

When the truck is on a level surface the cab should also appear level. To check the cab air suspension, make sure that the air-brake system pressure gauge reads the specified system cut-out pressure. Use a tape measure to check cab ride height at the location specified by the manufacturer. This is usually between the upper and lower plates. Compare your readings with specifications and adjust the leveling valve as necessary. Use a soap and water solution to check for leaks. Also check the condition of the shock absorbers in the cab suspension system by looking for fluid leaks and observing any roughness in the ride when the cab is at normal ride height. Also, check the condition of the transverse rod and mounts by looking for excess lateral movement.

Task B3.13 Inspect front bumper, fairings, and mounts.

Check the front bumper to make sure that it is securely mounted and is not cracked or otherwise damaged. Make sure that there is sufficient space between the ends of the bumper and the front tires to avoid damage to the tires. The fairings should also be checked for cracks and that the mounts are tight and undamaged.

4.   Air Conditioning and Heating (HVAC) (2 questions)

Task B4.1   Inspect A/C condenser and lines for condition and visible leaks; check mountings.

The technician/inspector must report and log all refrigerant leaks into a PMI schedule, so these repairs are addressed in a timely manner. All technicians that perform service on an A/C system of a motor vehicle must be certified, regardless of the capacity of the person performing the service (i.e., shop foreman, parts specialist, and technician).

The condenser is normally mounted just in front of the truck radiator, or if rooftop-mounted, in the center of the roof above the driver/passenger area. In either position, it receives the full flow of ram air from the movement of the truck.

Refrigerants carry oil with them and when they leak, there is usually an oily film around the source of the leak. Observe hoses, fittings, and seals for signs of leakage. Sometimes a dye has been added to the system to help locate leaks. These dyes are designed to be seen using a black-light leak detector. Smaller leaks may require the use of electronic leak detectors. Check the condensers mounting bolts and rubber insulators. A condenser that is allowed to move or is subject to vibration may lead to more serious leaks.

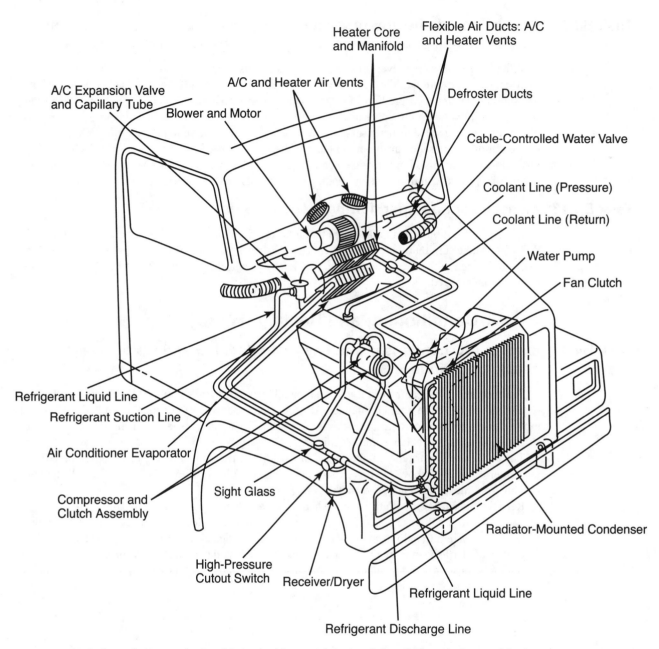

Location of components of a typical heavy-duty truck heating and air conditioning system.

Task B4.2   Inspect A/C compressor and lines for condition and visible leaks; check clutch; check mountings.

Depending upon the rate of leakage, the Clean Air Act establishes regulations concerning the repair of leaks. The technician/inspector must report and log all refrigerant leaks into a PMI schedule so these repairs are addressed in a timely manner (see the above figure).

Task B4.3   Check A/C system condition and operation.

Begin checking the A/C system condition and operation with a preliminary inspection. Check the compressor drive belt condition and tension. While checking the belt, look at the compressor mountings, checking for loose bolts or cracked mounts. Use a thickness gauge to check the compressor clutch condition. Visually check for damaged or leaking hoses and road debris buildup on the condenser coil fins. If the condenser fins are becoming plugged with debris, use compressed air, soapy water, a small brush, and a fin comb to clean the fins and restore airflow.

Engage the compressor with the engine idling and listen for indications of mechanical, pressure, or charge level problems. Some R-12 systems are equipped with a sight glass that may be used to check system charge, but R-134a systems usually do not. The use of a sight glass to check level of charge is not always conclusive and is being replaced by temperature testing by some technicians. Follow the OEM-recommended procedure for troubleshooting the system.

Frost on the receiver/dryer usually indicates an internal restriction in that component. Because the receiver/dryer is located in the high-pressure liquid line between the condenser and evaporator, it should feel warm. TXV valve frosting indicates a restricted valve or one sticking closed. Frost formation on the evaporator outlet indicates a flooded evaporator, caused by an excessive refrigerant charge, or a stuck-open TXV valve. Unusual noises can often guide the technician to a faulty component. These problems may also cause frost on the compressor suction line. On a system that uses a POA valve, frosting of the suction line is normal.

On systems using an accumulator, it should feel cold because of its close proximity to the evaporator. Both the evaporator inlet and outlet should feel cold when operating normally. On orifice tube systems, the evaporator inlet should be slightly warmer to the touch than the outlet. If the evaporator outlet is warm, the refrigerant charge may be low. High-side refrigerant components should feel hot or warm, and low-side components should feel cool or cold. A plugged drain in the evaporator case may cause a sulfurous odor in the cab. R-12 has no odor, while R-134a has a faint ether-like smell. A manifold gauge set is used to recycle, recharge, and diagnose A/C systems. The manifold gauge set is designed to access the refrigeration circuit and control refrigerant flow when adding or removing refrigerant. When the manifold test set is connected into the system, pressure is registered on both of the gauges at all times. Make certain that the hand shut-off valves are closed on the manifold before installation.

Task B4.4   Check HVAC air inlet filters and ducts; service as needed.

Change HVAC air inlet filters at least once a year, preferably at the beginning of summer. A clogged evaporator core drain will cause windshield fogging or a noticeable mist from the panel vents. A clogged drain normally can be opened with a slender piece of wire or with low-pressure shop air. The evaporator drain should be checked during routine maintenance inspections. A cracked evaporator case can cause a whistling noise during high blower operation. Minor cracks can be repaired using epoxy-type adhesives. A mildew smell that is noticeable during A/C system operation can be rectified by removing the evaporator case and washing it with a vinegar and water solution or a commercially available cleaner.

C. Electrical/Electronics (10 questions)

1. Battery and Starting Systems (4 questions)

Task C1.1   Inspect battery box(es), cover(s), and mountings.

An inspection of a battery should always begin with a thorough visual inspection of the case and the immediate surrounding area for any signs of leakage. Check mounting bolts, hold-down clips, and cover clamps. Rubber hold-downs and insulators should have no tears or cracks.

Task C1.2   Inspect battery hold downs, connections, cables, and cable routing; service as needed.

Battery terminals are either two tapered posts or threaded studs on top of the case or two internally threaded connectors on the side. Some newer batteries have both types of terminals so that one battery fits either application. These terminals connect to either end of the series of elements inside the battery and have either a positive or a negative marking, depending on which end of the series they represent. When disconnecting a battery, the ground cable should always be disconnected first, then the positive cable; therefore, when reconnecting a battery, the positive cable should always be connected first and the ground cable last. Heavy-duty trucks typically use three or four batteries connected in series or parallel to allow operation of a 12-volt or 24-volt electrical system, and/or to increase the CCA requirement of the vehicle. The most common hookup is the parallel hookup, which can be further described as a single-path circuit or dual-path circuit. The path is often determined by the placement of batteries, such as two batteries per side of truck or all the batteries in one battery box on one side of the truck or behind the cab.

Check the condition of the battery cables by performing voltage drop tests. Connect the voltmeter in parallel with the portion of the circuit being tested, being careful to observe polarity. For example, to test for resistance caused by battery terminal corrosion, place one voltmeter lead on the battery post, the other one on the cable, and then observe the meter while cranking the starter. A good connection at the battery post will show that only about 0.1 volt is dropped. When the voltage dropped across, the entire positive battery cable is measured there should be less than 0.5 volts dropped. You can test the positive battery cable by placing one lead on the battery post and the other on the solenoid battery cable stud, testing voltage drop through the entire cable. A similar test can be used to test the ground circuit: place one lead on the negative battery post and the other on the starter frame and observe the voltage reading while cranking. There should be less than 0.3 volt dropped across the negative cable. All wires somewhat resist current flow. The value of this resistance is relative to the size and type of wire used. In relation to current flowing through the wire, the smaller the wire is, the greater the resistance value. Knowing this, we can go one step further and realize that as the current value in amperes increases, the resistance value in ohms also increases due to the wrong size wire (too small). Resistance generates heat, which further increases resistance.

A battery can be cleaned with a baking soda and water solution. Always wear hand and eye protection when servicing batteries and their components. If the built-in hydrometer indicates light yellow or clear, the electrolyte level is low and the battery should be replaced. A high voltage regulator setting that causes overcharging may cause the low electrolyte level. When disconnecting battery cables, always disconnect the negative battery cable first.

It is a good practice to spray the cable clamps with a protective coating to prevent corrosion. A little grease or petroleum jelly will also prevent corrosion. Also available are protective pads that go under the clamp and around the terminal to prevent corrosion.

Task C1.3   Check and record battery state of charge (open circuit voltage) and condition.

The first step is a visual inspection of the battery case, terminals, and electrolyte level before testing can be performed. If there is any leakage from the case due to cracks or loose posts/terminals then the battery is deemed unserviceable and must be replaced. After this the level of the electrolyte (a combination of sulfuric acid and distilled water) must be serviced if required. The plate area must be completely immersed in electrolyte. If the level is low (indicated by the plate area being exposed), then distilled water only can be added to return the electrolyte to proper balance. Tap water contains impurities and will lead to premature battery failure.

A battery having moisture on the top surface, a strong odor of sulfur, expanded case or excessively high temperature usually indicates overcharging. Any moisture on the battery surface will cause a discharge condition and must be removed. Cold-weather starts are critical to the operation of a truck. A battery that is even partially discharged will be exposed to a freeze-up problem in cold weather. The specific gravity in a discharged battery is decreased to a point where the electrolyte can freeze.

A battery open-circuit test can be used to determine its state of charge. A battery that has recently been charged by running the engine, or using a battery charger, must have its surface charge removed before testing. Surface charge may be removed by applying a load of 300 amps for 15 seconds with a carbon pile load tester, or by cranking the engine for 15 seconds. After removing the surface charge, stabilize the voltage by allowing the battery to rest for 15 minutes. The following chart may be used to correlate voltage with state of charge.

State of Charge Determined by Open Circuit Voltage Test	
Stabilized Open Circuit Voltage	**State of Charge**
12.6 Volts or more	Fully Charged
12.4 Volts	75% Charged
12.2 Volts	50% Charged
12.0 Volts	25% Charged
11.7 Volts or less	Discharged

Task C1.4   Perform battery test (load and capacitance).

If the battery passes the state of charge test (Task C1.3) then perform a load test to determine if the battery is capable of providing the proper discharge amperage and voltage under load conditions.

Note: *A two-minute period should be maintained between any load testing cycles to avoid damage to the battery.* If this is a multiple-battery vehicle, then first disconnect the batteries and perform individual testing of each battery. Determine the amperage rating of the vehicle (check manufacturer's specifications) and confirm that the proper batteries are installed in the vehicle. If the voltage is below 12.4 volts DC at 70°F, the battery should be recharged to above 12.6 volts. Apply a load through the tester at one-half the cold cranking rating (CCA) for 15 seconds. If the voltage drops below 9.6 volts at 70°F, the battery is unserviceable. **Note:** *A quick test can be performed if the battery voltage indicates that it has to be recharged.*

As charging to proper levels can take hours to complete properly, it would be good time management to install the battery charger, set it to a medium charge rate (voltage), amperage approximately 40 amps, and observe the voltage rate. If the voltage exceeds 15 volts DC within two to three minutes then the battery requires replacement.

Note: *For safety and accuracy, testing cannot be performed on frozen batteries. Also, note that the sulphated battery test just mentioned may not be accurate on all battery types.*

Voltage/Temperature Table		
Minimum Voltage	**Temperature**	
	°F	**°C**
9.6	70	21
9.5	60	16
9.4	50	10
9.3	40	4
9.1	30	–1
8.9	20	–7
8.7	10	–12
8.5	0	–18

Task C1.5   Inspect starter, mounting, connections, cables, and cable routing.

Starter motors produce high torque, so check the clamping torque of the starter flange fasteners. The starter circuit carries the high-current flow within the system and supplies power for the actual engine cranking. Components of the starting circuit are the battery cables, magnetic switch or solenoid, and the starter motor. Starter relays may be integral or nonintegral with the starter. External starter relays (also known as a "mag" switch) can attach either to the starter as a separate unit or to the firewall. Removal of these types of starter relays entails removing the two positive leads and the control wires. The starter relay is an electrical high-current switch and, as such, it does not have a ground lead. If the starter system has a magnetic switch, jumper around the heavy terminals to see if the motor cranks. If it cranks, the relay is defective assuming that control current from the starting switch is available at the small terminal of the relay. Any time maintenance is to be performed on the starter, remove the ground (negative lead) from the battery to prevent accidental shorting of the wiring. A voltmeter is connected across a component in a circuit to measure the voltage drop across the component. A current must be flowing through the circuit during the voltage drop test. If a circuit has a normal resistance, the voltage drop is within specifications. For example, with the engine cranking, the positive battery cable may have a 0.2-volt drop. Excessive resistance in a cable or component causes a higher than specified voltage drop and reduced current flow. The sum of the voltage drops in any series circuit will equal the source voltage or applied voltage.

Task C1.6   Engage starter; check for unusual noises, starter drag, and starting difficulty.

While cranking the engine, listen for unusual sounds that could be indications of problems. If the starter is noisy while cranking the engine there may be loose or broken mounting bolts, or the teeth on the flywheel or starter drive may be worn or damaged. The starter may spin without cranking the motor, indicating a worn starter drive or damaged teeth. An engine that cranks slowly or unevenly can result in a no-start condition because a diesel engine has to reach a minimum cranking speed before it can start.

Whenever the engine cranks over too slowly remember that it could be caused by engine mechanical problems or it could be related to the starter. The starter may be receiving insufficient voltage or current from the battery. Other noises to listen for are a starter that chatters and disengages before the engine has started and a starter that engages but does not spin. Note when any of these noises are present, so that more thorough testing can be performed to find the cause of the problem.

High starter current draw, low cranking speed, and low cranking voltage usually indicate a faulty starter motor. Low current draw, low cranking speed, and high cranking voltage usually indicate excessive resistance in the starting circuit. The starting system circuit carries the high current flow within the system and supplies power for the actual engine cranking. The cranking current test or starter draw test measures the amount of current, in amperes, that the starter circuit draws to crank the engine. This amperage reading is useful in isolating the source of certain types of starter problems. An inoperative starter, along with a high current draw is always an indication of a direct short to ground. Any opens, failed switches, springs, or relays would cause an operative failure and a no-current condition.

2.   Charging System (4 questions)

Task C2.1   Inspect alternator, mountings, wiring, and wiring routing.

If the alternator mounting holes are distorted due to overtorquing and/or age, then the alternator should be replaced. Severely out-of-round mounting holes on an alternator will cause misalignment of the drive belt and may cause premature bearing failure. A complaint of an unusual noise requires a thorough visual inspection. A visual inspection will usually determine whether a loose component is causing a vibration or any unusual sounds. Drive belt tension also is critical to prevent slippage.

Task C2.2   Perform alternator current output test.

The minimum output requirement of the alternator (voltage and amperage) must first be confirmed to the manufacturer's specification. Determine if requirements will exceed output (i.e., due to additional lighting, trailer, refrigeration unit). Note: All testing is usually performed at 5,000 alternator rpm or 1,500 to 2,000 engine rpm for purposes of testing accuracy. If the vehicle is a late model and computer controlled, this speed may not be possible as the engine ECM will not permit anything above a preset limit when there is no load on the engine. Perform battery testing as outlined in Task C1.4 prior to this test. Using a load tester unit, place the positive cable of the tester onto the positive battery cable and the negative cables the negative battery cable. Then place the inductive pickup clamp of the tester around the output cable of the alternator. The load tester should now indicate battery voltage. Start the engine, and raise engine speed to the specified rpm. Apply a load using the carbon pile until the specified maximum amperage is read without dropping system voltage below 13 volts. Alternator output should be within 10 percent of the specified rating. As ambient temperature increases, alternator output may decrease. This is a normal condition.

Task C2.3   Perform alternator voltage output test.

This voltage test can be used in conjunction with Task C2.2 as the amperage output of the alternator will have an effect upon the voltage being produced. Information should be confirmed as to the manufacturer's specifications/settings. Voltage output of the alternator is to be measured (engine running) at the alternator output terminal using a voltmeter. Output will depend on load.

The in-dash type voltmeter is usually connected at or near the battery. To verify voltage accuracy of the dash voltmeter, connect the test meter on the battery terminals in proper polarity; the meter reading of the test and dash unit should be within a $\frac{1}{2}$ volt DC of the test unit.

3. Lighting System (2 questions)

Task C3.1   Check operation of interior lights; service as needed.

Check the operation of the interior lighting. Although inoperative interior lighting is not considered a basis for placing a vehicle out of service (OOS), it is still important for safe operation of the truck. The dome lights and luggage compartment lights are usually operated by a door jam switch that completes the ground circuit. When the door is opened, the light should come on. Some interior compartment lights use a mercury switch that closes when the door is opened.

Dash lighting should operate when the lights are on and the dimmer should adjust them to the desired level. Common lighting problems are power supply, ground, and bulbs. Making sure that the bulb is good, that there is adequate power, and that there is a good ground path will solve most problems. When more than one light has similar symptoms, such as not operating or being dim, look for what they have in common. For example, if all of the interior lights are dim, the cab may not be adequately grounded. This is especially true when lights are attached to plastic panels, and the ground must be completed elsewhere.

Continuity in an electric circuit may be tested with a high-impedance digital voltmeter. Connect the meter ground lead to a satisfactory ground connection and connect the positive meter lead to various locations in the circuit starting at the voltage source. When the meter reads a very low voltage, the open circuit is between the meter positive lead and the previous location where a normal voltage reading was obtained. A 12-volt test light may be used to test continuity in a circuit, but a high-impedance test light must be used when testing continuity in computer-controlled circuits. If an ohmmeter is used to test continuity, the meter leads must be connected across the terminals on the component to be tested with the component disconnected from the circuit. An ohmmeter may be damaged if it is connected to a circuit with voltage supplied to the circuit. Always disconnect the truck's battery before using a self-powered test lamp to test continuity. Never use a self-powered test lamp on an electronic module circuit, as damage to the module may result.

Task C3.2   **Check all exterior lights, lenses, and reflectors; check headlight alignment; service as needed.**

Exterior lights are generally mounted on the sides, roof, and front of the vehicle. On the sides (depending on cab design) are turn-signal lights, side-marker indicators, and sometimes an intermediate turn signal. On the roof are clearance lights (usually on either end), identification lights (usually in the middle), and sometimes additional utility lights.

In the front are single or dual headlights, fog lights, turn-signal lights, and sometimes additional utility lights.

In order to properly adjust the headlights, headlight aim must be checked first. Various types of headlight-aiming equipment are available. When using aiming equipment, follow the instructions provided by the equipment manufacturer. Where headlight-aiming equipment is not available, headlight aiming can be checked by projecting the upper beam of each light upon a screen or chart at a distance of 25 feet ahead of the headlights. The vehicle should be exactly perpendicular to the chart.

An operational check of the brakelight switch is performed during a check of the electrical system, since the brakelight or stoplight switch, as it is generally called, is part of the vehicle's electrical system.

Defective and/or improper ground connections are the most common sources of dim lights. In addition to replacing all defective lamps and bulbs, periodically check to see that all wiring connections are clean and tight, that lamp units are tightly mounted to provide a good ground and that headlights are properly adjusted.

Task C3.3   **Inspect and test trailer power cord connector, cable, and holder; service as needed.**

The junction block for the trailer lights is located inside the cab, often directly behind the driver seat. Access is usually gained by removing a plastic cover held in place by four screws. In sleeper models, the junction block is usually in the luggage compartment. Whenever a lighting inspection of a vehicle and/or combination vehicle(s) is to be performed, the inspector usually starts at the front and works back from there. This is because the source of power is usually up front.

In most modern box trailers, there are no splices for corrosion to attack and the trailer junction box has been eliminated. Most current vehicles use a seven-wire trailer plug and connector. Ensure that the prongs and sockets are not corroded.

D. Frame and Chassis (19 questions)

1. Air Brakes (6 questions)

Task D1.1   **Check parking brake operation.**

The parking brake must be capable of holding a fully loaded truck. A truck with a park brake that will not hold is dangerous, and should not be dispatched. Air brake equipped trucks will have park brakes on one or more of the nonsteering axles. They are applied by spring pressure and released by air pressure. Start the engine and wait until the air system reaches system pressure. Push in the yellow-colored, diamond-shaped dash control valve knob to release the park brakes. It should stay in; if it doesn't, check for a pressure loss or a defective control valve. Test the park brakes by putting the truck in low gear, moving it forward at idle speed, and applying the park brakes by pulling out on the knob. The truck should stop promptly. If it does not, repairs are needed. Park brake performance is affected by brake free-stroke adjustment. Excessive free-stroke will reduce both park and service brake performance.

Task D1.2   **Check and record air governor cut-in and cut-out settings (psi).**

The governor manages compressor loaded and unloaded cycles. It defines the system pressure. System pressure in most trucks is set at values between 120 and 135 psi, with 125 psi being most typical. System pressure is any value between cut-in and cut-out pressure. Cut-out pressure is the pressure at which the governor outputs the unloader signal to the compressor. The unloader signal is maintained until pressure in the supply tank drops to the cut-in value. Cut-in pressure is required by FMVSS 121 to be no more than 25 psi less than the cut-out value. The difference between the two values' range is between 20 and 25 psi.

Governor operation can be easily checked. One method is to drop the air pressure in the supply tank to below 60 psi and, with the vehicle's engine running, build up the pressure. A master gauge should be used to record the cut-out pressure value. This should be exactly at the specification value. If not, remove the dust boot at the top of the governor, release the lock-nut and turn the adjusting screw either clockwise to lower or counterclockwise to raise the cut-out pressure.

Governors seldom fail but they do not have check valves. If the unloader signal is not delivered to the compressor unloader assembly, high system pressures will result. If the safety pop-off valve on the supply tank trips (this occurs at 150 psi), this is usually an indication of governor or compressor unloader malfunction. The only adjustment on an air governor is the cut-out pressure value. If the difference between governor cut-out and cut-in is out of specification, the governor must be replaced.

Task D1.3   Service air drier as needed. Check air drier purge valve operation and air drier heater, if equipped.

Water and oil residue collected in the air dryer must be purged. Each time the compressor unloader cycles, it triggers the purge valve in the air dryer sump, ejecting moisture and any accumulated sludge. The purge valve may become clogged and need to be cleaned. During cold weather, it may freeze up if its heater fails. The heater should operate when the temperature falls below about 40°F. Test the operation of the purge valve by cycling the compressor through several cut-in/cut-out cycles, and observing the operation of the purge valve. Each time the unloader cycles, there should be a discharge of air and moisture from the bottom of the air dryer.

Task D1.4   Check air system for leaks (brakes released).

First, block the wheels on the vehicle. Build air pressure to governor cut-out, release the parking brakes and shut off the engine. Walk around the unit listening for audible air leaks. Then check dash gauges for system pressure loss. Investigate any out-of-specification loss. Leaks that cause prolonged pressure build-up are out-of-service items. According to the Commercial Vehicle Safety Alliance, a vehicle with more than 20 percent of its air brakes defective (measured with the engine off, reservoir pressure at 80–90 psi, and the brakes fully applied) shall be declared out of service.

Task D1.5   Check air system for leaks (brakes applied).

Check the air system for leaks with the brakes fully applied. Begin by starting the engine and bringing the air supply to system pressure. With the engine idling, make and hold a full brake application until pressure stabilizes. If the compressor is unable to maintain reservoir pressure of 80–90 psi with the engine at idle and the brakes fully applied, it is recommended that the truck be deadlined until repairs are completed. Shut off the engine, and have an assistant hold the brakes applied while you examine the truck for audible air leaks.

Task D1.6   Test one-way and double-check valves.

The supply tank feeds the primary and secondary reservoirs of the brake system. Each is pressure protected by means of a one-way check valve. One-way check valve operation can be verified by draining the supply tank and watching for pressure loss in the other reservoirs. One-way check valves are used variously throughout an air brake system to pressure-protect and isolate portions of the circuit.

Two-way (double) check valves play an important role as a safeguard in dual circuit air brake systems. The typical two-way check valve is a T with two inlets and a single outlet. It outputs the larger of the two source pressures to the outlet port and blocks the lower pressure source; the valve will shuttle in case of a change in the source pressure value. In other words, it will always prioritize the higher source pressure. Two-way check valves provide a means of providing the primary circuit with secondary circuit air and vice versa in case of a circuit failure. In the event that both source pressures are equal, as would be the case in a properly functioning dual air brake circuit, the valve will prioritize the first source to act on it. This would usually be from the primary circuit.

Task D1.7   Check low air pressure warning devices.

Before performing a pressure drop test on a heavy truck a technician should ensure the engine is stopped and the system is at full pressure. Then make and maintain a brake application. A block of wood may be used to hold the brake pedal down during these tests. Allow pressure to stabilize for one minute, and then begin timing for two minutes while watching the dash gauge. Pressure drop for single vehicles is 4 psi within two minutes and for singles' combination vehicles 6 psi within two minutes.

System air build-up times are defined by federal legislation, specifically FMVSS 121. This legislation defines the required build-up times and values. A common check performed by enforcement agencies requires that the supply circuit on a vehicle be capable of raising air system pressure from 85–100 psi in 25 seconds or less. Failure to achieve this build-up time indicates a worn compressor, defective compressor unloader assembly, supply circuit leakage, or a defective governor.

FMVSS 121 requires that a driver receive a visible alert when the system pressure drops below 60 psi; in most cases, this is accompanied by an audible alert, usually a buzzer. A low air pressure warning device is fitted to both the primary and secondary circuits. This is a simple electrical switch that can be plumbed anywhere into a system requiring monitoring. The switch is electrically closed whenever the air pressure being monitored is below 60 psi. When the air pressure value exceeds 60 psi, the switch opens.

Verifying the operation of a low air pressure warning switch can be done by pumping the service application valve until the system pressure drops to the trigger value. A dash-located gauge must monitor both the primary and secondary circuit air pressures. The required visible warning is usually a dash warning light. An optional audible alert is almost always used in conjunction with the visible alert.

Task D1.8   Check spring brake inversion/emergency (spring) brake control valve, if equipped.

To check for proper functioning of the inversion (spring brake control) valve, the air system should be fully charged (minimum 100 psi), and the spring parking brakes must be released. Next, locate and drain the wet tank and then the primary axle service reservoir completely by means of the draincocks. The primary gauge needle will drop to 0 psi, indicating a pressure loss for that system, which is normal. The secondary gauge needle must remain at system pressure (minimum 100 psi) to start the check of the inversion valve. Depress the foot valve in a normal manner. The spring brakes on the rear axle(s) will apply as the air is exhausted from the brake air chambers. Release the foot valve, and the spring brakes will release as air is supplied to them from the secondary reservoir. If the spring brakes do not or only partially release, then the inversion valve has failed and should be replaced. Remember, an inversion valve is a failsafe device and does not operate when brakes are functioning properly.

Task D1.9   Check tractor protection valve.

The tractor protection valve protects the tractor air supply under a trailer breakaway condition or severe air leakage. Two air lines connect a tractor brake system with a trailer air circuit. The trailer supply line supplies the trailer with air for braking and any other pneumatic systems such as its suspension. The trailer service line is the service brake signal line. Both these lines are plumbed to the tractor protection valve. When the trailer supply dash valve (emergency) red hexagal is in its off position, the tractor protection valve is also off. When the driver makes a service brake application, no air will exit the tractor protection valve service signal line to the trailer. When the trailer air supply valve is open, air will open the tractor protection valve and supply the trailer pneumatic systems with air. Then when the driver makes a service brake application, the service brake signal air will be transmitted through the tractor protection valve to apply its service brakes.

It should be noted that the trailer air supply controls the trailer park brakes. Whenever air supply is interrupted, such as when the trailer is being parked or unintentionally in a breakaway, the spring brakes will apply regardless of how much pressure is in the trailer air reservoirs.

The tractor protection valve is designed to isolate the tractor air system from that of the trailer at a predetermined value that ranges from 20 to 45 psi, depending on the system. It should be noted that at these pressures, the spring brakes on both the tractor and trailer would be partially applied.

Task D1.10   Test air pressure build-up time.

System air build-up times are defined by federal legislation, specifically FMVSS 121. This legislation defines the required buildup times and values. A common check performed by enforcement agencies requires that the supply circuit on a vehicle be capable of raising air system pressure from 85–100 psi in 25 seconds or less. Failure to achieve this build-up time indicates a worn compressor, defective compressor unloader assembly, supply circuit leakage, or a defective governor.

Task D1.11   Check condition and operation of hand brake (trailer) control valve, if equipped.

The trailer control valve is used to actuate the trailer service brakes on a trailer independently of the tractor service brakes. This valve is also known by the terms trolley valve, broker brake, and spike. The source of air supplied to the trailer control valve is usually the secondary circuit. Brake application pressure air is modulated proportionally with control valve travel.

Two-way check valves are used in the application circuit of the trailer brakes. The two-way check valve is located downstream from both the trailer control valve and the treadle valve. In a typical system, the trailer application brake valve (trolley or spike) uses secondary circuit sourced air and the treadle valve trailer service signal uses primary circuit sourced air to activate the trailer service brakes. If a driver was using the spike to bring the vehicle to a stop and during this process an emergency occurred that required a panic application of the treadle valve, the two-way check valve would permit the source air to change from secondary circuit air from the trailer control valve to the higher source pressure value from the treadle valve.

Trailer control valve performance can be verified with an accurate air pressure gauge. The trailer is supplied with air by means of air hoses. Service, and trailer supply, air is sent to the tractor protection valve. This air is then transferred to the trailer by hoses with couplers known as gladhands. Gladhands enable easy coupling between the tractor and the trailer pneumatic circuit.

Task D1.12   Perform antilock brake system (ABS) operational system self-test. Perform automatic traction control (ATC) operational system self-test, if equipped.

All current ABSs are electronically managed. A typical ABS adapts a standard air or hydraulic system for electronic management. All current systems are designed to default to nonmanaged operation in the event of an electrical or electronic malfunction. New trucks are equipped with ABS brakes and are mandatory.

An ABS consists of a means of monitoring the speed of each wheel, an ECU or ABS module to manage the system, and a modulator assembly to affect the outcomes of the ECU logic processing by modulating the pressures to each wheel. ABS brakes are especially effective at managing split-coefficient braking conditions. Split-coefficient braking occurs when road surface conditions (icy, wet, gravel, dry, etc.) differ from one wheel to another.

All electronic ABS are equipped with self-diagnostics. These can be accessed by flash or blink codes using the dash warning light or a scan tool (digital diagnostic reader) connected either at the ATA connector in the cab or directly to the ABS module. Each time an electronic ABS is powered up, a self-test is performed. This consists of testing the signal range in the wheel speed sensors and cycling the solenoids in the system actuators. The latter test is audible, producing a clicking noise.

Access the ABS self-diagnostics by selecting the correct message identifier (MID). Identify the failure mode indicator (FMI) that has triggered the fault code and repair the failure using the OEM service instructions.

Task D1.13 Inspect coupling air lines, holders, and gladhands.

Visually check all hoses and lines in the brake system. Service and trailer supply air is plumbed from the tractor protection valve. This air is transferred to the trailer by air hoses with couplers known as gladhands. Gladhands enable coupling between the tractor air supply and the trailer pneumatic circuit. Gladhand seals are usually manufactured out of rubber. They are retained in an annular groove in the gladhand and are easily replaced when they fail.

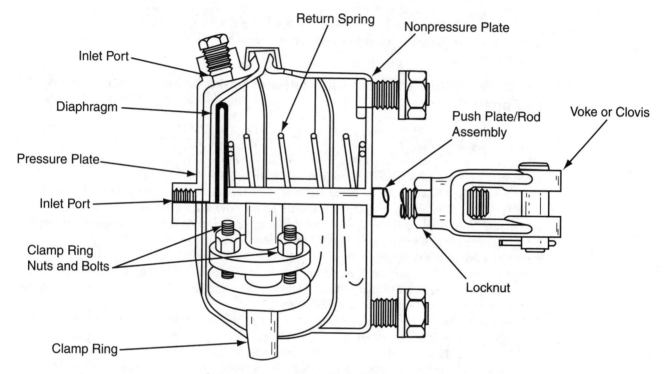

Cutaway view of a typical air brake chamber.

Task D1.14 Check brake chambers and air lines for secure mountings, damage, and missing caging plugs.

Air-brake chambers should be checked for signs of damage (see figure above). Visually check for damage from road hazards and look for cracks around the mounting studs. Check the air chamber bracket for cracks. Any hose or tubing that is worn, pinched, abraded, leaks, or swells under pressure should be replaced. Air lines should be routed to avoid damage from moving parts, and secured so that they are not damaged by an accumulation of ice during freezing weather. Here are some of deficiencies that should be noted during an annual inspection:

- Absence of braking action on any axle, required to have brakes, upon application of the service brakes.

- Missing or broken mechanical components including: shoes, lining pads, springs, anchor pins, spiders, cam rollers, push rods, and air chamber mounting bolts.

- Loose brake components including air chambers, spiders, and camshaft support brackets.

- Audible air leak at brake chamber (ruptured diaphragm, loose chamber clamp, etc.).

- Hose with any damage extending through the outer reinforcement ply. Rubber impregnated fabric cover is not a reinforcement ply. Thermoplastic nylon may have braid reinforcement or color difference between cover and inner tube. Exposure of second color is cause for rejection.

- Bulge or swelling when air pressure is applied.

- Any audible leaks from hose or tubing.

- Two hoses improperly joined (such as a splice made by sliding the hose ends over a piece of tubing and clamping the hose to the tube).

- Air hose cracked, broken, or crimped.

- Tubing cracked, damaged by heat, broken, or crimped.

Task D1.15 Inspect and record front and rear brake lining/pad condition and thickness.

Brake linings that have any of the following conditions should be replaced.

- Lining or pad is not firmly attached to the shoe

- Saturated with oil, grease, or brake fluid

- Nonsteering axles: Lining with a thickness less than ¼ inches at the shoe center for air drum brakes, ¹⁄₁₆ inches or less at the shoe center for hydraulic and electric drum brakes, and less than ⅛ inches for air disc brakes.

- Steering axles: Lining with a thickness less than ¼ inches at the shoe center for drum brakes, less than ⅛ inches for air disc brakes, and ¹⁄₁₆ inches or less for hydraulic disc and electric brakes.

Unequal lining wear should be noted and the cause investigated.

Truck brake shoes are in most cases fixed-anchor assemblies mounted to the axle spider and actuated by an S-camshaft. Brake shoes for a heavy-duty truck can have their friction facings or linings mounted to the shoe by bonding, riveting, or by screw fasteners. Current application shoes are remanufactured and replaced as an assembly, that is, with the new friction facing already installed. When reusing shoes, check for arc deformities usually caused by prolonged operation without brake adjustment. The lining blocks are tapered and seldom require machine arcing.

Brake shoes must be fitted with the correct linings. Lining requirements can change with different vehicles and applications. When reconditioning shoes with bolted linings, if the original fasteners are reused, the lock washers should at least be replaced. Riveted linings should be riveted in the manufacturer's recommended sequence.

The friction rating of linings is coded by letter codes. Combination lining sets of shoes are occasionally used; these use different friction ratings on the primary and secondary shoes. When combination friction lining sets are used, care should be taken to install the lining blocks in the correct locations on the brake shoes.

It is good practice to replace all of the brake linings on a tandem drive axle truck on a PMI schedule. When the linings on a single wheel are damaged, as in the case of wet seal axle lube failure, the linings of both wheel assemblies on the axle should be replaced to maintain brake balance.

The friction pads in air disc brake assemblies should also be changed in paired sets, that is, both wheels on an axle. Vented disc brakes may have thicker inner pads than outer pads, while solid disc brakes usually have equal thickness inner and outer pads. Heat transfers more uniformly in the solid disc assemblies than vented ones, requiring thicker inboard pads.

Task D1.16 Inspect condition of front and rear brake drums/rotors.

Brake drum condition can vary according to many different factors, including but not limited to, scoring, bell-mouth condition, concave-like, convex-like, threaded appearance, and hard spots.

For example, convex condition indicates a drum diameter greater at the friction surface edges as compared to the center of the friction surface. Concave condition indicates a drum diameter greater at the center of the friction surfaces as compared to the edges. A bell-mouthed condition occurs when the drum diameter is greater at the inboard edge of the drum (next to the backing plate) compared to the outboard edge. The out-of-round condition is a variation in the drum diameter at measurements taken 180 degrees apart.

When checking for lateral run out, a dial indicator is usually mounted on the hub and the face of the rotor is indicated. Total indicated run out should generally not be more than 0.005 inch (0.13 mm).

If brake drums are to be reused, they should be inspected for wear, heat discoloration, scoring, glazing, threading, concaving, convexing, bellmouthing, and heat checking. It is not usually possible to machine a used brake drum that is warped due to heat tempering (recent practice and the low cost of drums usually results in drum replacement with a brake job). It is good practice to machine new drums before installation due to warpage caused by incorrect storage practices after manufacture. The maximum legal service limit for 14-inch-diameter brake drums is 14.120 inches. The machining limit is 14.060 inches and the discard limit is 14.090 inches. A drum micrometer is used to check brake drum diameter.

Brake rotors must be visually inspected for heat checking, scoring, and cracks. One must measure rotors for thickness with a micrometer. You use a dial indicator to check rotor run out and parallelism. In highway applications, rotors tend to outlast drums and are often reused after a brake job. They must be turned within legal service specifications using a heavy-duty rotor lathe.

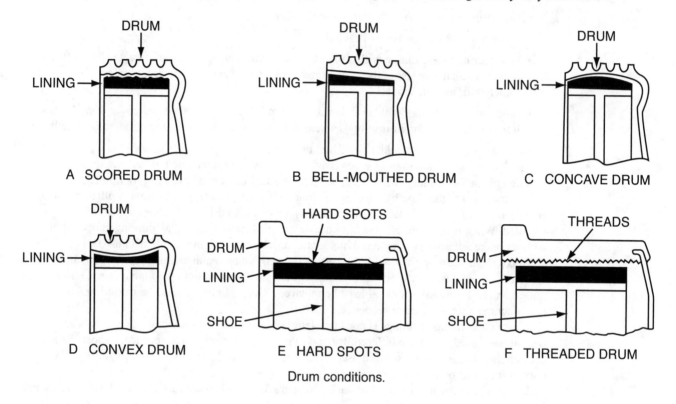

Drum conditions.

Type	Outside Diameter	Maximum Stroke at Which Brakes Must Be Readjusted
6	4½	1¼
9	5¼	1⅜
12	5¹¹⁄₁₆	1⅜
16	6⅜	1¾
20	6²⁵⁄₃₂	1¾
24	7⁷⁄₃₂	1¾
30	8³⁄₃₂	2
36	9	2¼

"LONG STROKE" CLAMP TYPE BRAKE CHAMBER DATA

Type	Outside Diameter	Maximum Stroke at Which Brakes Must Be Readjusted
16	6⅜	2.0
20	6²⁵⁄₃₂	2.0
24	7⁷⁄₃₂	2.0
24*	7⁷⁄₃₂	2.5
30	8³⁄₃₂	2.5

BOLT TYPE BRAKE CHAMBER DATA

Type	Outside Diameter	Maximum Stroke at Which Brakes Must Be Readjusted
A	6¹⁵⁄₁₆	1⅜
B	9³⁄₁₆	1¾
C	8¹⁄₁₆	1¾
D	5¼	1¼
E	6³⁄₁₆	1⅜
F	11	2¼
G	9⅞	2

BOLT TYPE BRAKE CHAMBER DATA

Type	Outside Diameter	Maximum Stroke at Which Brakes Must Be Readjusted
9	4⁹⁄₃₂	1½
12	4¹³⁄₁₆	1½
16	5¹³⁄₃₂	2
20	5¹⁵⁄₁₆	2
24	6¹³⁄₃₂	2
30	7¹⁄₁₆	2¼
36	7⅝	2¾
50	8⅞	3

WEDGE BRAKE DATA
Movement of the scribe mark on the lining shall not exceed ¹⁄₁₆ inch.

*For 3 in. maximum stroke-type 24 chambers

Task D1.17 Check operation and adjustment of front and rear brake automatic slack adjusters.

The slack adjuster is critical in maintaining the required free-play and adjustment angle. Current slack adjusters are required to be automatically adjusting, but their operation must be verified routinely. Manually adjusting automatic slack adjusters is dangerous and should not be done, except during installation or in an emergency to move the vehicle to a repair facility, because manual adjustment of this brake component (1) fails to address the true reason why the brakes are not maintaining adjustment, giving the operator a false sense of security about the effectiveness of the brakes, which are likely to go out of adjustment again soon; and (2) causes abnormal wear to the internal adjusting mechanism for most automatic slack adjusters, which may lead to failure of this brake component.

Slack adjusters connect the brake chambers with the foundation assemblies on each wheel. They are connected to the pushrod of the brake chamber by means of a clevis yoke (threaded to the pushrod) and pin. Slack adjusters are spline mounted to the S-cams and positioned by shims and an external snap ring.

The slack adjuster converts the linear force of the brake chamber rod into rotary force or torque and multiplies it. The distance between the slack adjuster S-cam axis and the clevis pin axis will define the leverage factor; the greater this distance, the greater the leverage.

The objective of brake adjustment, whether manual or automatic, is to maintain a specified drum to lining clearance and a specified amount of free-play. Free-play is the amount of slack adjuster stroke that occurs before the linings contact the drum. Grease or automatic lubing systems lubricate slack adjusters. The seals in slack adjusters are always installed with the lip angle facing outward. When grease is pumped into the slack adjuster, grease will easily exit the seal lip when the internal lubrication circuit has been charged.

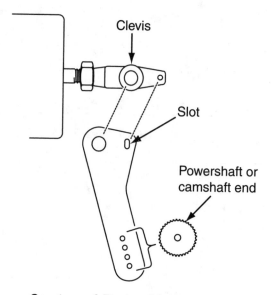

Courtesy of Rockwell International.

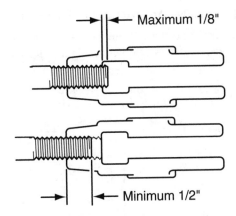

Courtesy of Rockwell International.

Failure to adjust the slack adjuster clevis (see figure above) with the proper template causes improper self-adjusting action in the slack adjuster. Observe the outer clevis pin hole through the slot in the template. If necessary, adjust the clevis on the end of the pushrod until the outer hole in the clevis is completely visible through the template opening. With the clevis properly adjusted, the pushrod must extend at least $1/2$ inch (1.27 mm) into the clevis threads, and not more than $1/8$ inch (3.17 mm) into the yoke (see figure above).

Rotate the adjusting nut on the automatic slack adjuster so the openings in the slack adjuster arm are aligned with the openings in the clevis. Place anti-seize compound on the two clevis pins and insert these pins through the clevis and slack adjuster arm. When replacing a failed slack adjuster, the distance between the axis of the S-cam bore and the clevis pin bore must be maintained—a difference of $1/2$ inch can greatly alter the brake torque and unbalance the brakes.

Brake Chamber Free Stroke Measurement

Follow this procedure to perform the free stroke measurement.

- Measure the distance from the brake chamber face to the center of the large clevis pin with the brakes released. Record this measurement.

- Insert a pry bar in the slack adjuster clevis and extend the brake chamber pushrod until the shoes contact the drum. Measure the distance between the center of the large clevis pin and the brake chamber face. The difference between these two measurements is the free stroke.

- If the free stroke is not within specifications adjust the slack adjuster (adjusting) nut to obtain the specified free stroke.

Brake Chamber Applied Stroke Measurement

- Measure the distance from the center of the large clevis pin to the brake chamber with the brakes released (see following figure). Record this measurement.

Measure "Free Stroke"

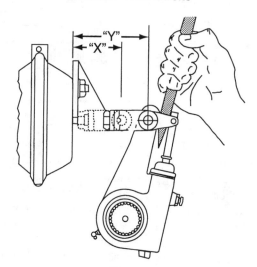

Free stroke = "Y" minus "X"
Drum brake free stroke must be 1/2–5/8 inches.
(12.7–15.9 mm)
Disc brake free stroke must be 3/4–7/8 inches.
(19.1–22.2 mm)

- Have a co-worker apply the brakes until the application pressure is 80 to 90 psi (551.6 to 620.55 kPa). Measure the distance between the center of the large clevis pin and the brake chamber face. The difference between these two measurements is the applied stroke (see following figure).

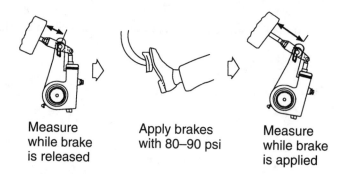

Measure
while brake
is released

Apply brakes
with 80–90 psi

Measure
while brake
is applied

- If the applied stroke is not within specifications, adjust the slack adjuster (adjusting) nut to obtain the specified applied stroke.

Task D1.18 Check camshaft and bushing condition.

When performing foundation brake servicing, all of the hardware mounted on the axle spider should be inspected and replaced if required. There should be no radial play of the S-camshaft. Rollers and cams should be inspected for flat spots, and it is good practice to replace any spring-steel components such as retraction springs and snap rings. Spider fastener integrity should be checked at each brake job. Spiders should be inspected for cracks at each brake job.

S-cam bushing seals should be installed so that the spider outboard bushing sealing lip faces inward toward the bushing and the inboard bushing sealing lip faces outward from the bushing. This allows excess grease to exit from the inboard side of the bushing and prevent it from being pumped into the foundation assembly. S-cam profiles depend on friction and should never be lubricated.

Task D1.19   Lubricate all air brake component grease fittings.

If a slack adjuster passes inspection, the following procedure for lubrication should be followed. Apply a thin film of chassis grease to the slack-adjuster splines. After reassembly, pressure-lubricate the slack adjuster according to the manufacturer's instructions. Pressure lubricate the in-board S-cam bushing until grease flows out of the slack adjuster end of the tube. Grease should not flow out of the out-board end of the tube. If it does, the seal is defective and it must be replaced. Also, do not replace the existing grease fitting(s) with the pressure relief type. Only standard nonvented fittings are to be used with spring loaded lip seals.

2.   Hydraulic Brakes (2 questions)

Task D2.1   Check master cylinder for leaks and damage; check fluid level and condition.

When checking hydraulic brake fluid in a vehicle, not only should the level be checked, but any leaks from the master cylinder, reservoir, or brake line fittings as well. The appearance of the fluid is also an important part of the inspection. A cloudy or opaque look to the fluid will indicate some sort of contamination (dirt, rust, etc.), especially if severe. Fluid that is cloudy or darkened should be flushed and refilled with fresh fluid.

Task D2.2   Inspect brake lines, fittings, flexible hoses, and valves for leaks and damage.

A proper inspection of a heavy truck hydraulic braking system should include a visual checking around the wheel cylinders for an accumulation of contamination, which generally indicates leakage. Brake lines should be in good condition with no corrosion or damage that may compromise system performance. A thorough inspection of brake lines is always considered an important part in inspecting the hydraulic brake system.

Task D2.3   Check parking brake operation, inspect parking brake application and holding devices.

A check of the parking brake controls should come under an inspection of the hydraulic brake system. Test actuator free-play and the parking brakes ability to effectively stop the vehicle. See figure below.

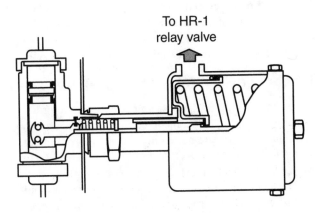

To HR-1
relay valve

Task D2.4   Check operation of hydraulic system; pedal travel, pedal effort, and pedal feel (drift).

The brake pedal should feel firm with minimal free travel at the top of the stroke. If the pedal feels spongy, it is usually an indication of air in the hydraulic system. Trucks with hydraulic brakes manufactured after 1976 are equipped with a tandem master cylinder, and dual circuit brakes (see figure). In a dual circuit brake system, when one circuit fails, the remaining circuit will be unaffected. When there is a failure in one circuit, the brake pedal will have excessive free travel. A dash warning light should also be illuminated.

If the pedal slowly drifts to the floor there may be an internal leak in the master cylinder, or an external fluid leak. When the pedal is depressed, the brakes should apply promptly without excessive effort. If excessive pedal effort is required to stop the truck, the power brake booster should be checked. When the pedal is released, the brakes should release promptly. When the brakes are applied, all of the wheels should brake evenly and the truck should not pull to either side. If there is a pull, or one wheel locks up, or skids, the problem should be located and repaired before the truck is dispatched.

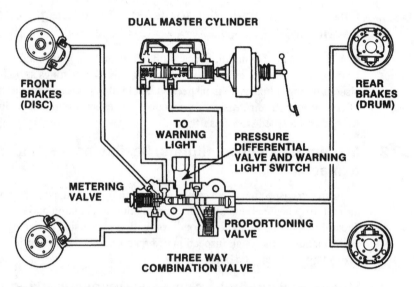

Schematic of a dual circuit, split disc/drum hydraulic brake system.

Task D2.5   Inspect wheel cylinders/calipers for leakage and damage.

A proper inspection of a heavy truck's hydraulic braking system should always include a visual checking around the wheel cylinders for an accumulation of contamination, which generally indicates leakage. Look for evidence of leaks on the inboard side of the tire and wheel. When the drums are removed as part of a PMI, visually check the wheel cylinders by peeling back the rubber boot and checking for fluid. Leaking calipers may be examined with the wheel removed. Leakage from wheel cylinders can result in brake failure. Leaking wheel cylinders must be reconditioned or replaced.

Task D2.6   Inspect power brake booster(s), hoses, and control valves.

Most trucks are equipped with power brakes to provide shorter stopping distances with less effort. A vacuum booster, a pneumatic booster, or a hydraulic booster may provide power brake assist. Hydraulic boosters are more effective and more common. When road testing the truck, it should stop without excess pedal effort. There should also be reserve capacity, providing for at least one power-assisted brake application after the engine is off.

The hydro-max power brake booster accomplishes this by engaging an electro-hydraulic pump when there is an interruption of fluid flow from the engine-driven hydraulic pump. The hydraulic power source may be the power steering pump, or may be a dedicated hydraulic pump. Vacuum assisted systems have a vacuum reservoir that provides reserve capacity. Trucks with diesel engines and vacuum boosters need an engine driven vacuum pump, because a diesel engine has pressure, not vacuum in the intake manifold. Air-over-hydraulic systems use air pressure for power assist, and the air in the tanks provides reserve capacity. Inspect the hydraulic system for fluid leaks. Check the hoses for signs of leaks, and rubber deterioration.

There are a number of control valves required for braking balance with disc/drum combinations. The metering valve delays application of front disc brakes until the rear drum brakes are ready to apply. If only the front brakes lock up or skid on slick surfaces when the brakes are lightly applied, the metering valve should be checked. You can test the metering valve by installing pressure gauges on the front and rear brake circuits. There should be no pressure at the front until the pressure at the rear reaches about 30–50 psi.

A proportioning valve is needed to reduce rear wheel lock-up during moderate to heavy brake applications. Some trucks are equipped with load sensing proportioning valves, which allow more pressure to the rear wheels as the frame is lowered by heavier loads. The rear wheels should not lock up before the front. There should be balanced braking at each of the trucks wheels. A pressure differential valve turns on a dash warning light when a loss of pressure occurs in one half of a split hydraulic brake system. Combination valves combine the metering, proportioning, and pressure differential valves into one assembly, which is replaced as a unit.

Task D2.7   Inspect and record front and rear brake lining/pad condition and thickness.

Most disc brake assemblies today have wear indicators, which produce a noise when the wear limit is exceeded. Servicing disc brake pads usually involves removing the caliper assembly, ensuring that float pins are not seized, backing off the automatic adjusting mechanism, and installing a new pair of brake pads. When a self-adjusting mechanism is used, care should be taken to ensure it is properly activated on reassembly. Whenever the brake pads are replaced, the brake rotors should be both measured and visually inspected. It is good PMI practice to replace linings and pads that are close to their wear limit.

When a brake job is performed, the brake shoes, return springs, and fastening hardware should be replaced. The friction face codes should be observed when replacing brake shoes. Most brake shoes today use bonded friction blocks, but riveted and bolted types are still in existence. Ensure that primary (leading) and secondary shoes are installed in their correct locations.

Task D2.8   Inspect condition of front and rear brake drums/rotors.

Brake drum condition can vary according to many different factors including, but not limited to, scoring, bell-mouth condition, concave-like, convex-like, threaded appearance, and hard spots. The drum in the figure indicates a convex-type condition.

Brake drums may be reused if they are within the manufacturer's specifications. The critical specifications are the maximum wear limit, machine limit, and maximum permissible diameter. Drums are measured with a drum gauge and should be checked for out-of-round, bell-mouthing, convexing, concaving, and taper. Before machining drums, they should be inspected after measuring for heat checks and cracks. Disc brake rotors may be reused if they are within the manufacturer's specifications. They should be measured for thickness with a micrometer and checked for parallelism and runout with a dial indicator. If within machine limits, the rotor may be turned on a rotor lathe.

Task D2.9   Adjust drum brakes.

Adjust drum brakes as follows:

- Remove the backing plugs for measuring the shoe-to-drum clearance. Be sure that, if used, the parking brake chambers are caged.

- Apply and release the brake pedal several times to center the brake shoes.

- Insert a feeler gauge through the backing plate opening and measure the clearance between the brake shoe and the drum. Typically, this clearance should be .040 to .100 inch (1.01 to 2.54 mm) (see figure).

- If the shoe clearance is not correct, turn the manual adjusting wheel (star wheel) on the wheel cylinder to obtain the correct clearance.

Task D2.10   Check operation of hydraulic assist back-up system and warning devices.

Trucks equipped with hydraulic-assist power brakes (usually hydraulic pressure is supplied from the power steering pump) require back-up to supplement power braking in the event the engine stops while the truck is moving. This is accomplished by a flow switch, which senses loss of hydraulic pressure from the power steering pump and actuates a 12-volt electro-hydraulic pump to provide emergency power assist. On some systems this is tested by making several brake applications with the key on and the engine off. When you have made enough pedal applications to exhaust the pressure in the accumulator, you should get a warning light and buzzer, and hear the 12-volt electro-hydraulic pump engage. Other systems are controlled by a solid state alarm/brake booster module and can verify when they are operating normally by observing both a primary and a secondary brake light and a tone alarm. This system is necessary for safety. It should be checked during each inspection.

3.   Drive train (3 questions)

Task D3.1   Check operation of clutch brake.

The clutch brake is a circular disc with a friction surface that is mounted on the transmission input spline shaft between the release bearing and the transmission. Its purpose is to slow or stop the transmission input shaft from rotating in order to allow gears to be engaged without clashing and to keep transmission gear damage to a minimum. Clutch brakes are used only on vehicles with nonsynchronized transmissions. Worn clutch brake friction surfaces can lead to clashing (grinding) when moving the gear lever into reverse or low gear while vehicle is stationary.

To check the operation of the clutch brake, fully depress the clutch pedal in the cab and note the point of resistance that occurs before the end of the pedal stroke. Clutch brake squeeze should measure 1 inch at the clutch pedal. Most manufacturers recommend special tools for setting clutch brake. Check that the clutch brake friction faces are grease-free and the drive tabs are free.

Task D3.2   Check clutch linkage/cable and levers for looseness or binding.

When you service the release bearing on the clutch, all pivot points on the clutch linkage should be checked for free movement, inspected, and lubricated. On a push-type clutch, adjust clutch-pedal free travel. On pull-type clutches with a clutch brake, adjustment must accommodate proper free travel and clutch brake settings. Note that pull-type clutch internal adjustments must be completed prior to any external linkage adjustment. External linkage adjustments are usually not necessary.

Task D3.3   Check clutch master cylinder fluid level. Check clutch master cylinder, slave cylinder, lines, and hoses for leaks and damage.

Trucks equipped with a hydraulic clutch master and slave cylinder arrangement should have the master cylinder fluid level checked each time it is serviced. When fluid must be frequently added, it indicates that there is a leak in the system. Most leaks are found with a visual inspection of the cylinders, lines, and fittings. Brake fluid is most often used in clutch hydraulic systems. Hydraulic clutch master cylinders use DOT 3 brake fluid. The rubber seals in the hydraulic clutch system are incompatible with other fluids such as motor oil, power steering fluid, and hydraulic oils. Clutch master cylinders are usually not repairable and must be replaced.

Task D3.4   Check clutch adjustment.

Apply the clutch pedal. Free travel should measure between 1 and 2 inches, with $1\frac{1}{2}$ inches being typical. Measure the distance the pedal can be pushed down before thrust is exerted on the clutch release bearing to check clutch-pedal free travel.

Riding the clutch pedal means operating the vehicle with the clutch partially engaged. This can be destructive to the clutch as it permits slippage and generates excessive heat. Riding the clutch also puts constant thrust load on the release bearing, which can thin out the lubricant and cause excessive wear on the pads. Release bearing failures are often the result of this driving practice.

Both push- and pull-type clutches are disengaged through the movement of a release bearing. The release bearing is a unit within the clutch that mounts on the transmission input shaft but does not rotate with it. A fork attached to the clutch-pedal linkage controls the movement of the release bearing. As the release bearing moves, it forces the pressure plate away from the clutch disc assembly.

Manually adjusted clutches have an adjusting ring that permits the clutch to be adjusted to compensate for friction lining wear. The ring is positioned behind the pressure plate and threaded into the clutch cover. A lock plate secures the ring so that it cannot move. Levers are seated in the ring. When the lock strap is removed, the adjusting ring is rotated in the cover so that it moves toward the flywheel.

The wear compensator is a replaceable component that automatically adjusts for facing wear each time the clutch is actuated. Once facing wear exceeds a predetermined amount, the wear compensator allows the adjusting ring to be advanced toward the flywheel, maintaining pressure plate to clutch disc clearance within proper operating specification. This also keeps free pedal adjustment within specification.

The following five-step procedure can be used to determine if adjustment is required on manual and self-adjusting clutches with a clutch brake.

1. Remove the inspection cover.

2. With the clutch engaged (pedal up), measure the clearance between the release bearing housing and the clutch brake. This is the release travel. If the clearance is less than $\frac{1}{2}$ inch or greater than $\frac{9}{16}$ inch, continue with steps 3 and 4; if less than $\frac{1}{2}$ inch or greater than $\frac{9}{16}$ inch, go to step 5.

3. Release the clutch (pedal down) by fully depressing the clutch pedal.

4. Using the internal adjustment procedures for lock-plate and auto-adjust mechanisms, advance the adjusting ring until a distance of $\frac{1}{2}$ inch to $\frac{9}{16}$ inch is attained between the release bearing housing and the clutch brake (clutch cover) with the clutch pedal released (pedal down).

5. If the clearance between the release bearing housing and the clutch brake (clutch cover) is less than specifications, rotate the adjusting ring counterclockwise to move the release bearing toward the flywheel.

Task D3.5   Check transmission(s) case, seals, filter, cooler, and cooler lines for cracks and leaks.

The transmission case must be checked for any signs of fatigue. Cracking is a symptom that is usually accompanied by fluid leakage. Inspect all external connection points for evidence of leakage.

Task D3.6   Check transmission wiring, connectors, seals, and harnesses for damage and proper routing.

The transmission wiring is essential to the operation of the truck. Other chassis computerized systems depend upon the transmissions tailshaft sensor for road speed data. Many new transmissions have computer controls, which are multiplexed with the rest of the truck electronics. A careful examination of the transmission's electrical components is important when performing preventive maintenance. Transmission wiring should be routed and secured so moving parts or hot exhaust pipes do not damage it. Connectors should be checked for external damage. Avoid disconnecting them each inspection because they are designed to be weatherproof, and frequent disconnections can damage them. Make sure they are clean and dry.

Task D3.7   Inspect transmission(s) breather(s).

Plugged breathers are associated with fluid leakage but not at the location of the breather. When a breather becomes plugged, fluid is often forced past seals or gaskets in the transmission. If a technician jumps to a conclusion when they notice fluid leakage at a seal, they may replace the seal and think the problem is solved. The technician should check the transmission breathers during any transmission diagnosis and PMI.

Task D3.8   Inspect transmission(s) mounts for looseness and deterioration.

Transmission mounts and insulators play an important role in preventing drive train vibration from transferring to the chassis of the vehicle. If the vibration were allowed to transmit to the chassis of the vehicle, the life of the vehicle would be reduced. Driving comfort is another reason why insulators are used in transmissions. Insulators absorb shock and torque. If the transmission was mounted directly to a rigid frame, the driveline torque required to haul heavy loads would need to be absorbed by the transmission resulting in shorter service life. Broken transmission mounts are not readily identifiable by any specific symptoms. They should be visually inspected during each PMI.

Task D3.9   Check transmission(s) oil/fluid level and condition.

Check transmission oil by removing the fill plug; oil should be exactly at the level of the fill hole or slightly above. During a PMI, fill the transmission to the correct level exactly. Do not overfill. This can cause aeration.

Cold temperature operations can cause transmission problems. High-viscosity oils typically used in transmission and axles thicken and may not flow at all when the vehicle is first started. Rotating gears can push lubricant aside, leaving voids or channels where no lubricant is actually contacting the gears. This is referred to as channeling. It is not until heat generated from under-lubricated gear contact surfaces melts the gelled lube that it starts to flow and do its job.

Task D3.10   Inspect U-joints, yokes, drive lines, and center bearings for looseness, damage, and correct phasing.

Loose end yokes, excessive radial runout, slip spline slop, bent shaft tubing, or missing plugs in the slip yoke cause vibrations, U-joint, and center bearing problems. A PMI will normally include the inspection of all of these, plus checking the center bearing for damage and the trunnions of the U-joints for wear. When inspecting drivelines, also check the driveshaft(s).

Often, vibration is too quickly attributed to the driveshaft. Before condemning the driveshaft as the cause of vibration, the vehicle should be thoroughly road-tested to isolate the vibration cause. To assist in finding the source, ask the operator to determine what, where, and when the vibration is encountered. It is very helpful to keep in mind some of the causes of driveline vibration. U-joints are the most common source if the vibration is coming from the driveshaft, while driveshaft balancing is the next most common. Pay special attention to phasing when removing or installing a driveshaft.

Center support bearings are used when the distance between the transmission and the rear axle is too great to span with a single driveshaft. The center bearing is fastened to the frame and aligns the two connecting driveshafts. It consists of a stamped steel bracket used to align and fasten the bearing to the frame. A rubber mount inside the bracket surrounds the bearing, which supports the driveshaft. The truck manufacturer's recommended driveshaft rating should always be adhered to when replacing a driveshaft. Oversized or undersized driveline components will lead to frequent driveline failures. Vibrations at high speeds are often indicative of a driveline problem.

Task D3.11   Inspect axle housing(s) for cracks and leaks.

A technician should be able to identify potential drive axle problems. Incorrect lubrication will greatly affect the life of bearings, gears, and thrust washers. A lubrication problem will affect, and often cause, fatigue on drive train components including the drive axles.

Identify the root cause of any external leakage. Clean the housing thoroughly to remove accumulations of grime and dirt. Plug the breather and any other openings and steam clean or pressure-wash the axle assembly. Visually inspect for loose or missing fasteners, for stripped cross-threaded holes and for cracks in the housing. Look for cracks wherever there is a weld, such as around the flange where the spider is attached, and where the center housing is attached. Pay particular attention to suspension attaching points and spring saddle mounts, because these are high stress areas. Cracks can be identified with dye and UV light.

Task D3.12 Inspect axle breather(s).

Investigate any source of leaks found on the axle differential. Replacing a seal if the axle differential breather is plugged does not cure leaking seals. A plugged breather produces high axle housing pressure, which can result in leakage past seals. Clean the breather by soaking in solvent and blowing it out with compressed air each time the lube level is checked.

Task D3.13 Lubricate all drive train grease fittings.

Overgreasing can be as damaging as undergreasing. Grease guns exert tremendous pressure and oozing may mean that a bearing seal has blown out. The slip joint should be greased until grease is forced out of the relief hole. Then the hole should be covered until grease begins to ooze out around the seal. Clean zerk nipples before greasing to avoid contamination. Lubricate the cross and bearings of U-joints until the lube appears at all four seals. This flushes abrasive contaminants from each bearing assembly and assures all four are purged.

Task D3.14 Check drive axle(s) oil level.

Drive axle oil level is checked in the same way transmission oil level is checked. The oil should be exactly at the level of the fill hole or slightly over. Never overfill: this can cause aeration. Ensure that EP (extreme pressure) lubricant is used in units using hypoid gearing. Try not to mix lubricants with different specifications.

Task D3.15 Change drive axle(s) oil and filter; check and clean magnetic plugs.

A technician can sometimes determine the condition of a drive axle differential by the fluid. Most drive axles are equipped with a magnetic plug that is designed to attract metal particles suspended in the gear oil. A nominal amount of "glitter" is normal, but excess "glitter" indicates a problem requiring further investigation.

When metal particles are found on the magnetic drain plug, the technician should inform the customer of the condition and instruct the customer to monitor the condition and change oil in 30 days for reevaluation in case more serious problems are developing in the drive axle.

Drain and flush the factory-fill axle lubricant of a new or reconditioned axle after the first 1,000 miles (621 km) and never later than 3,000 miles (1864 km). This is necessary to remove fine particles of wear material generated during break-in that would cause accelerated wear on gears and bearings if not removed. Subsequent lube changes should be made at 100,000-mile intervals for linehaul operation, and at 40,000-mile intervals, or annually, for other operations.

Draining the lubricant while the unit is still warm ensures that any contaminants are still suspended in the lubricant. Flush the axle with clean axle lubricant of the same viscosity as used in service. Do not flush axles with solvents such as kerosene. Avoid mixing lubricants of a different viscosity or oils made by different manufacturers. Ensure that differentials with hypoid gearing are filled with an EP rated lubricant.

Task D3.16 Check two-speed axle unit operation and oil level.

Although some vehicles are equipped with electrical shift units, most two-speed axles are equipped with pneumatic shift systems. There are two air-activated shift system designs predominantly used to select the range of a dual range tandem axle or to engage a differential lockout. Usually the air shift unit is not serviceable. If it is found defective it should be replaced. Check the oil level at oil fill hole; it should be level with it.

Task D3.17 Change transmission oil/fluid and filters; check and clean magnetic plugs.

Most manufacturers suggest a specific grade and type of transmission oil, heavy-duty engine oil, or straight mineral oil, depending on the ambient air temperature during operation. Do not use multipurpose gear oil when operating temperatures are above 230°F (110°C). Many of these gear oils break down above 230°F (110°C) and coat seals, bearings, and gear with deposits that might cause premature failures. If these deposits are observed (especially on seal areas where they can cause oil leakage), change to heavy-duty engine oil or mineral gear oil to assure maximum component life.

Always follow the manufacturer's exact hydraulic fluid specifications. For example, several transmission manufacturers recommend DEXRON, DEXRON II, and type C-3 ATD-approved SAE 10W or SAE 30 oils for their automatic transmissions. Type C-3 fluids are the only fluids usually approved for use in off-highway applications. Type C-3 SAE 30 is specified for all applications where the ambient temperature is consistently above 86°F (30°C). Some, but not all, DEXRON II fluids also qualify as type C-3 fluids. If type C-3 fluids must be used, be sure all materials used in tubes, hoses, external filters, seals, etc., are C-3 compatible.

Task D3.18 Take transmission(s) oil sample.

Drain transmission oil. Halfway through the drain-off, fill the sample container. Correctly label and send for analysis.

Task D3.19 Take drive axle(s) oil sample.

During a PMI, transmission and drive axle oil samples are taken for analysis of lubricating qualities, metal, and dirt content. Also, corrosion-causing agents such as sulfuric and hydrochloric acids are tested. For this reason, the transmission/drive axle oil should be at normal operating temperature to ensure that all possible contaminants and corrosives be contained in the oil in its normal operating condition.

4. Suspension and Steering Systems (3 questions)

Task D4.1 Check steering wheel operation for free play or binding.

The best method for checking a steering system is with the wheels on the floor and the engine running. Allow the engine to idle with the transmission in neutral and the parking brakes applied. While someone turns the steering wheel one-quarter turn in each direction from the straight ahead position, observe all the pivoted joints on the tie-rod ends and drag link and other steering systems-related components. This allows the technician to check the steering linkage pivots under load. Turn the steering wheel back and forth while visually checking for looseness. Manual steering wheel play should not exceed 2 inches and power steering wheel play should not exceed 2½ inches.

A loose steering gear mount could be binding the steering column and cause the power steering to become increasingly harder to turn. Elongated mounting holes in the power steering pump bracket may cause a noise while in operation. Worn holes in the power steering pump-mounting bracket could cause premature belt wear. If leaks occur at any of the seal locations, replace the seal. When a leak is present at the high-pressure fitting, first tighten it to the specified torque. If this fails, replace the O-ring at this fitting.

Task D4.2 Check power steering pump and hoses for leaks and mounting; check fluid level.

Reservoir O-rings, driveshaft seals, high-pressure fittings, and dipstick caps are all possible leak sources. If leaks occur at any of the seal locations, replace the seal. When a leak is present at the high-pressure fitting, tighten it to the specified torque. If this fails, replace the O-ring or if necessary, the complete fitting. Check fluid level at operating temperature of about 175°F.

Task D4.3   Change power steering fluid and filter.

Most OEMs recommend the use of engine oil or automatic transmission fluid in power steering systems. Low fluid level will cause increased steering effort and erratic steering. It may also cause a growling or cavitation noise in the pump. Foaming in the remote reservoir may indicate air in the power steering system. Most OEM truck manufacturers recommend checking the power steering fluid level at operating or working temperature of 175°F.

With the engine at 1,000 rpm or less turn the steering wheel slowly from stop-to-stop several times to raise the fluid temperature. Check the reservoir for foaming as a sign of aerated fluid. The fluid level in the reservoir should be at the hot full mark on the dipstick. If the power steering fluid is contaminated with metal filings, improper fluid, or moisture, the system must be flushed. After the system is flushed, the filter in the remote reservoir must be replaced.

Task D4.4   Inspect steering gear(s) for leaks and mounting.

The steering gear must be mounted true to the frame or binding will result. Any evidence of leakage at the steering gear requires removal and repair.

Task D4.5   Inspect steering shaft U-joints for condition and phasing; inspect pinch bolts, splines, Pitman arm-to-steering sector shaft, drag link, and tie rod ends.

Steering systems use U-joints to allow the shafts to turn on an angle. Loose steering-wheel to steering-shaft fasteners can cause excessive play and poor steering response.

Do not compress a column steering U-joint with a C-clamp or large pliers when testing a linkage joint for wear. This can compress the tension spring in the joint and give the impression that there is wear in the joint. Push against the joint with hand force. This should be sufficient to identify excessive wear.

A loose Pitman arm on the steering gear output shaft could cause steering shimmy. The Pitman arm is responsible for directional stability. A damaged Pitman arm can account for the steering wheel being off center. A loose sleeve clamp on the drag link adjuster may allow enough play in the sleeve to damage the adjusting threads on the steering arm, enough to change steering wheel position. Tie-rod clamp interference with the I-beam on a full turn can cause a noise when turning over bumps, a damaged tie-rod assembly, and inaccurate camber measurement. A misadjusted drag link could cause the steering gear to operate off center and cause steering shimmy.

Task D4.6   Check kingpin and thrust bearing wear.

Use the procedure indicated below for checking kingpin fit for play. A dial indicator is mounted as shown, and the top of the wheel is moved in and out. The indicator reads bushing wear and kingpin slop. Then reposition the dial indicator so that up-and-down movement will be indicated and test the thrust bearing by moving the tire up and down and watching the movement of the steering knuckle.

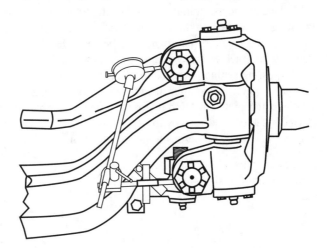

Task D4.7   Check front and rear wheel bearings for looseness and noise.

Always observe TMC wheel end procedure when adjusting wheel bearings. Bearing adjustment is very important on trucks equipped with ABS to ensure the correct wheel speed sensor air gap. To check bearing endplay, mount a dial indicator so that the indicator tip is contacting the hubcap-mounting surface, push the hub inward, zero the indicator, and pull the hub outward. Wheel bearing endplay should be 0.001 inch to 0.005 inch.

Defective rear wheel bearings produce growling noises at low speeds. A growling noise produced by a defective front wheel bearing is most noticeable while turning a corner. Rear wheel bearings usually do not require repacking because most are lubricated with rear-axle fluid. Never spin a dry wheel bearing with compressed air because it can damage critical contact surfaces.

Task D4.8   Check oil level and condition in all non-drive axle hubs; check for leaks.

Hubs should be prelubed when installing the wheel assembly. After installation, the bearing and hub assembly must be filled to the indicated level in the calibrated inspection cover. Ensure that the rubber boot cap seals properly and that there are no cracks in the plastic calibration plate.

Task D4.9   Remove and inspect front wheel bearings; reassemble and adjust.

Wheel bearings should be inspected at each brake job and at each PMI service. Current trucks use wet bearings, that is, they are lubricated with liquid lubricant, usually gear lube.

Bearings should be cleaned with solvent and air dried, ensuring that the cone is not spun out by the compressed air. The bearing should be inspected for spalling, galling, scoring, heat discoloration, and any sign of surface failure. When reinstalling wet bearings, they should be prelubed with the same oil to be used in the axle hub. When grease-packed bearings are used, the bearing cone must be packed with grease. This procedure may be performed by hand or using a grease gun and cone packer. Grease-lubricated wheel bearings require the use of a high-temperature axle grease. It is bad practice to pack wet bearings with axle grease and it may shorten bearing life by reducing lubrication efficiency.

Bearings fail primarily because of dirt contamination. This may be due to the conditions a vehicle is operated under or poor service practice. Lubrication failures caused by an inappropriate lubricant or lack of lubricant (caused by a failed wheel seal) also account for bearing failures; lack of lubricant will result in rapid failure and a bearing welded to an axle.

Maladjusted bearings fail rapidly, especially when preloaded. The result of high preload on a bearing can be to friction-weld it to the axle. Overloaded bearings tend to fail over time, not rapidly.

Most highway heavy-duty trucks use taper roller wheel bearings. The correct method of adjusting a taper roller wheel bearing is that outlined by the Truck Maintenance Council (TMC) division of the American Trucking Association (ATA). All bearing manufacturers in North America have endorsed this method, which supercedes any previous adjustment procedures outlined by them. The TMC bearing adjustment method requires that the bearing be seated to a specified torque value at the adjusting nut while rotating the wheel bearing; this is designed to seat the bearing with a preload. Next, the adjusting should be backed off to between one-sixth and one-third of a turn to locate the adjusting nut to the jam mechanism. Finally, the bearing end play must be measured with a dial indicator. The required specification is between 0.001 and 0.005 inches. End play must be present. If at least 0.001-inch end play is not present, the adjustment procedure should be repeated.

The bearing setting is locked in place on the axle spindle by a lock or jam nut that should be torqued to specification after the adjustment procedure. In some axles, a split forged or castellated nut and cotter pin are used to retain the wheel assembly. Current recommended practice in the truck service industry is defined by the TMC and the above method was developed to establish a single standard to counter numerous highway wheel-off incidents.

Task D4.10 **Inspect front and rear suspension components (springs, hangers, shackles, spring U-bolts, insulators, radius rods, torque rods, load pads, walking beams, and equalizers). Retorque U-bolts in accordance with manufacturers' specifications.**

The leaf spring type suspension uses semi-elliptic leaf springs to cushion load and road shocks. The springs are mounted on saddle assemblies above the equalizer beams and are pivoted at the front end on spring pins and brackets. The rear ends of the springs have no rigid attachment to the spring brackets, but are free to move forward and backward to compensate for spring deflection. Springs should be inspected for cracks, sag, and surface wear. Multiple leaf spring assemblies should be replaced as an assembly.

Task D4.11 **Inspect shock absorbers for leaks and mounting.**

When a vehicle hits a bump, the wheel and suspension move upward in relation to the chassis. This causes jounce and rebound. Shock absorbers dampen spring action from jounce and rebound, reduce body sway, and improve directional stability and driver comfort. Worn-out shock absorbers will allow the front end to bounce, causing steering wheel shake. Shock absorbers dampen spring oscillation and are always required on air spring suspensions. It is advisable to replace shock absorbers in pairs. Inspect shock absorbers with one-end disconnected, for resistance, bent pistons, bushing failure, and bent piston sleeves.

Task D4.12 **Inspect air suspension components (springs, mounts, arms, hoses, valves, linkage, and fittings) for leaks and damage.**

Air spring suspensions provide a smoother ride and automatically adjust ride height. A height control (leveling) valve controls the air supply to the air springs to maintain a constant ride height. These systems may have single- or dual-height control valves. The control and relay valves provide a means of deflating the air springs when uncoupling a trailer from a tractor. If the air springs were blown, the suspension would ride low. Misadjusted leveling (height control) valves in the rear air suspension system will cause a too-high condition. The height control valve lever should be parallel to the ground in the neutral position when the air suspension has been properly inflated to level the vehicle.

In a rear-axle air suspension system the air springs are mounted between the frame and the main suspension support beams and these beams are attached to the rear axle.

Some trailer air suspension systems have a rigid beam on each side of the suspension. The front of this beam is retained in a bracket with a bushing. These systems use an eccentric pivot bolt in this bracket for axle alignment.

Task D4.13 **Check suspension ride height.**

All suspension types have height specifications. Check the manufacturer's specifications regarding specific adjustment and measuring procedures. Usually spring types are measured with the vehicle not having any load installed (empty). With air spring suspension it usually will not make a difference as the height should not change loaded or not. Ride height is very important as it will impact on center of gravity, handling and driveline angles. Cab-over models require setting the suspension before clutch adjustment is attempted.

Most manufacturers recommend settings be made by deflating the air system and measuring the specification on the upward stroke of the fill cycle. It is also recommended to drive vehicle slowly onto a level shop floor, that brakes not be applied, chock wheels, and allow for suspension fore and aft movement while performing adjustments. Have air system pressure at minimum 100 psi. Confirm measurement height and place of measurement on the vehicle. Road test and reconfirm.

Task D4.14 Lubricate all suspension and steering grease fittings.

Lubricant specified by the manufacturer for the type of application and service interval must be used. Most manufacturers recommend that shop air pressure be regulated when applying grease to components. The normal suggested setting is 15 psi. This setting is recommended because the grease is required to purge the rust and moisture. If high pressure is used, then the grease merely bypasses the debris and does not provide a cleaning action. Steering gear shafts commonly use universal joints. These also require periodic lubrication. Binding of the steering wheel while turning could be caused by binding universal joints in the steering shaft. Steering not returning to center could be caused by lack of lubricant on the pivot points.

Automatic chassis lube systems (ACLS) eliminate the need for hand lubrication and increase component life. An automatic chassis lube system consists of the following components. There must be a reservoir to hold the grease. An automatic timer regulates how often grease is distributed. This can be adjusted to meet operating conditions. There may be an air-driven pump actuated by a solenoid valve, or an electric motor-driven pump. Air-driven pumps use metering devices to dispense the correct amount of lube to each component. Other systems use an air or electric motor driven pump with multiple outlets that meter the correct amount of lube at the pump. Many current vehicles are now equipped with at least some permanently lubricated components. These do not have grease fittings installed.

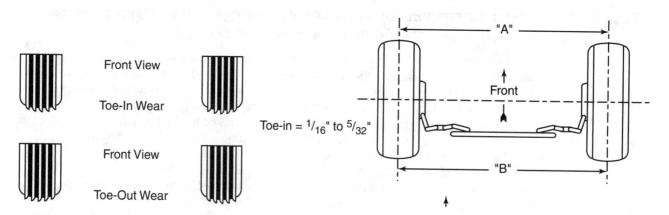

Note: Overhead View

Front View — Toe-In Wear

Front View — Toe-Out Wear

Toe-in = $^1/_{16}$" to $^5/_{32}$"

"A" "B" Front

Task D4.15 Check tandem axle spacing.

It is not necessary to replace the axle-alignment washers when replacing a radius rod on a truck. If the washers are not worn beyond use, then they should be reused to ensure a correct realignment of the axle. Also, the number of washers and their respective positions (forward or rearward) should be noted and returned exactly as removed.

A proper directional thrustline occurs when a vehicle's rear wheels track directly behind the front wheels. This prevents dog tracking or abnormal Ackerman effect. Worn walking (equalizing) beam bushings can cause the rear axles to be out of alignment. Replacing a bent radius rod could correct axle alignment. An axle shaft replacement will not correct the alignment of the axle assembly. A defective tracking bar could cause an axle alignment problem. Turning the adjustment eccentric is a method of adjusting the rear-axle alignment. Rotating eccentric bushings in the spring hangers does adjust the rear position on some spring designs.

5. Tires and Wheels (3 questions)

Task D5.1 Inspect tires for irregular wear patterns and proper mounting of directional tires.

Tires should be inspected when a "walkaround" is performed. Rub your hand across the tread to feel for irregular wear. The number and inboard/outboard location of heavy truck tires makes it imperative that a separate check of tires be made exclusive to a road test. Mismatched tires affect the drivability and stability of a vehicle; mismatching tires can cause handling and braking problems that seriously compromise safety. Many tires are unidirectional; check the tread geometry to ensure they are properly mounted.

Task D5.2   Inspect tires for cuts, cracks, bulges, and sidewall damage.

Cuts in the tread and sidewall should be noted for safety purposes. Tires with cuts that show exposed tire cords must be replaced. Cracking can be caused by exposure to the harmful rays of the sun. Bulges in the tread area can indicate a broken belt below the tread surface (road test might indicate steering wheel side-to-side movement at low speeds or vibration at highway speeds). Sidewall damage (bulges, cuts) can lead to future tire failure.

Task D5.3   Inspect valve caps and stems.

Perform a visual inspection and observe damaged, bent, or misaligned valve stems. Valve caps must be installed to prevent loss of air pressure. Salt, sand, and ice can cause air in the valve stem to discharge.

Task D5.4   Measure and record tire tread depth; probe for imbedded debris.

Measure the tread depth at the inboard, center, and outboard tread grooves, and at several places around the circumference of the tire. Tires on the steer axle must have at least $4/32$ inch tread remaining at the most worn spot. Tires on drive, trailer, and dolly axles must have at least $2/32$ inch tread remaining. Look for small rocks and debris imbedded in the tread and remove any found. If left, they may burrow into the casing and cause a leak.

If any repair bulge is $3/8$ inch above the tire surface, the tire must be replaced. If any tire has bumps, bulges, or cracks indicating separation of the tire plies, replace the tire. If retread tires have an opening at the edge of the tread, insert a probe in this opening. If the probe can be inserted more than a $1/2$ inch into the opening, replace the tire.

Task D5.5   Check and record air pressure.

Periodic checking and record keeping of air pressure is the most effective means of prolonging tire wear. Check manufacturer service bulletins for recommended up-to-date air pressure settings. Air contained within the tire is what supports the entire weight of the vehicle. Stability, directional control, and braking are all affected by tire pressure.

Task D5.6   Check for loose lugs and/or slipped wheels; check mounting hardware condition; service as needed.

Wheel nuts and lugs must be correctly torqued. Over- or under-torquing of stud-piloted wheels can cause broken studs and cracked or loose wheels. The single flange nuts of hub-piloted wheels are less susceptible to this problem. Slipped wheels caused by loose lugs can be observed on split rim design if the paint on the rim is scuffed as the wheel drags during braking and rust marks emerge. Mounting hardware must be of the proper grade to provide the clamping and tensile strength necessary to hold wheels/rims in place and have the ability to stretch or bend without breaking. If any mounting studs are stretched or broken, then all on that wheel location must be replaced.

Task D5.7   Retorque lugs/nuts in accordance with manufacturers' specifications.

At recommended intervals retorquing is necessary. This is due to the constant flexing of the stud/hub material. Overloading can be a factor. Manufacturers have specified intervals recommended for wheel and rim inspections.

Task D5.8   Inspect wheels and spacers for cracks or damage.

Periodic maintenance intervals are important for the inspection of wheels and spacers. A visual inspection can reveal cracks in the rim area. To check for cracking in the wheel nut mounting area, sometimes the nut will have to be removed as some designs cover the mounting hole. Cracked wheels are usually caused by vehicle overloading. Spacers are designed to keep dual wheels from having the tires contact each other causing friction/overheat, chafing, etc. They must be correctly installed to keep wheels in alignment. Generally, spoke wheels experience greater alignment and balance problems than disc designs, but with proper installation and torquing, it is possible that spoke wheels can run virtually trouble-free.

Task D5.9   Check tire matching (diameter and tread) on dual tire installations.

Tires of a different tread design mounted on the same axle are a deadline condition in many jurisdictions. You cannot mix radial tires with bias tires on a steering axle. Mixing tires on the rear axle is dangerous because radial tires expand more than bias tires. Matching tire sizes on dual wheels prevents tire tread wear from slippage due to uneven surface areas. Dual drive wheels may be measured with a tape measure, or caliper. Low-profile radial truck tires enhance the radial design to produce even lower costs per mile. The low-profile name comes from the tire's aspect ratio, which, for any tire, is calculated by dividing the tire's section height (tread center to bead plane) by its section width (sidewall to sidewall). Simply put, low-profile tires are "squatter" than conventional radials. Advantages offered by low-profile radials include lower weight (up to 10 percent less than standard radials), lower rolling resistance (again about 10 percent less), greater vehicle stability due to a lower center of gravity, a better footprint as a result of improved pressure distribution, high retreadability, improved fuel economy, better traction, and increased tread life.

6.   Frame and 5th Wheel (2 questions)

Task D6.1   Inspect fifth wheel mounting bolts, air lines, locks, and stops.

The fifth wheel (see following figure) is the standard semi-oscillating type that articulates on an axis perpendicular to the vehicle centerline. This design gives excellent stability and versatility while minimizing weight and space. The fifth wheel is bolted to the tractor frame by means of Grade 8 fasteners. Fastener integrity and torque should be checked at each PMI. The maximum allowable fifth wheel height is determined by subtracting the trailer height and tractor frame height from the maximum height of 13 feet, 6 inches. Many trailers are built with a 47-inch (119.38 cm) kingpin height. However, the trailer kingpin height must be measured to be sure of this height. The technician must remember the fifth wheel is designed to operate with the trailer level.

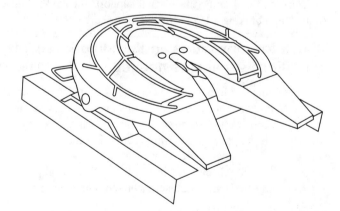

Task D6.2   Test operation of fifth wheel locking device.

The fifth wheel couples the tractor to the trailer kingpin. During the tractor-to-trailer coupling process, the fifth wheel should be adjusted so the trailer bolster plate makes initial contact with the fifth wheel at a point 8 inches (20.32 cm) to the rear of the center on the fifth wheel mounting bracket. After the tractor is coupled to the trailer, the driver should get under the tractor and trailer and use a flashlight to visually inspect that the fifth wheel jaws are locked.

When checking fifth wheel adjustment, movement should not exceed ¹/₈-inch horizontal play at any time. The jaws should also be checked for ease of operation and the sliders should work smoothly. A sliding fifth wheel is designed to move forward or rearward on its mounting plate. This type of fifth wheel is mounted on tracks and locked in position. The locking mechanism may be released mechanically with a lever or by air pressure supplied to an air cylinder. Test the operation of either type.

Task D6.3   Check mud flaps and brackets.

These are required to be present. Check for damage and tears; replace if necessary.

Task D6.4   Check pintle hook assembly, mounting bolts, and locks.

A draw bar is used to pull dollies and full trailers. A large steel ring on the front of the draw bar is coupled to a pintle hook on the back of the lead trailer. The pintle hook is bolted at the centerline of the lead trailer. A pintle hook has a spring-loaded latch that prevents the draw bar from disengaging. The pintle hook must be carefully examined to prevent accidents. Make sure the latch is in good condition and is securely mounted. Check for signs of wear or cracks. CVSA out-of-service criteria calls for deadlining if any welded repairs are visible, frame cross members are cracked in the attachment area, fasteners are missing or ineffective, or cracks are present anywhere in the pintle hook or draw bar assembly. No part of the draw bar eye should have any section reduced by more than 20 percent.

Task D6.5   Lubricate fifth wheel plate and all grease fittings.

Grease should be applied to the pivots and smeared over the deck. Fifth wheel lubrication is very important to maintain proper fifth wheel life and proper operation. A dry fifth wheel top surface may cause jerky steering. The top surface of the fifth wheel must be lubricated with a water-resistant lithium-based grease. Lubrication of the fifth wheel top plate may be done by direct application or through the grease fittings on the underside of the top plate. To distribute grease, most automatic systems use the same components. These components are as follows: a reservoir, which holds the grease supply until ready to be distributed; a pump, which delivers the lubricant through a network of grease lines; metering valves, which dispense grease; automatic timer mechanisms, which direct the flow of grease; and electric motors and/or other power sources (such as compressed air that pump it).

Task D6.6   Inspect frame and frame members for cracks and damage.

To measure a truck frame, the vehicle should be parked on level ground, such as on a smooth floor or pavement. Measure the frame angle with a bubble protractor or other leveling device. "Down in the rear" indicates a positive frame angle. "Up in the rear" indicates a negative frame angle.

Frame sideways occurs when one or both frame rails are bent inward or outward. The following are some causes: collision damage, fire damage, and using the truck for other than original design.

Frame sag occurs when the frame rails are bent downward in relation to the rail ends. The following are some causes:

- Excessive loads
- Uneven weight distribution
- Holes drilled in frame flanges
- Too many holes drilled in the frame web
- Holes drilled too close together in the frame web
- Welding on the frame
- Cutting holes in the frame with a torch
- Cutting notches in the frame rails
- Fire
- Collision damage
- Using the truck for other than original design

Frame buckle occurs when one or both frame rails are bent upward in relation to the ends of the rails. The following can cause buckle: collision, using the truck for other than design intent, and fire.

Diamond-shaped frame occurs when one frame rail is pushed rearward in relation to the opposite frame rail. Vehicle tracking is affected by this condition. The following can result in diamond frame: collision, towing, or being towed with the chain only attached to one side of the truck frame.

Frame twist occurs when the end of one frame rail is bent upward or downward in relation to the opposite frame rail. The following causes can result in a twisted frame frame: collision damage, rollover, rough terrain operation, and uneven loading.

When fabricating frame reinforcements, taper the reinforcement plates to avoid abrupt changes in section modulus. Each reinforcement plate edge should have an angle less than 45 degrees.

Task D6.7   Inspect body attaching hardware.

If the truck is equipped with a body such as a refrigerated or dry cargo body it will be attached to the frame with brackets, clamps, and bolts. The attaching hardware between the truck frame and the body should be checked during each inspection. Look for loose bolts, cracked brackets, and shiny places along the frame indicating that the body may be shifting.

Task D6.8   Inspect cargo ramps. Inspect lift gates, cylinders, controls, hoses, and wiring. Check fluid level.

Cargo ramps and lift gates are essential to safely moving freight. They should receive careful attention like all other safety devices. Cargo ramps should be checked for signs of cracking and fatigue. Lift gates should operate dependably and safely. Check the hydraulic pumps, cylinders, and hoses for signs of leakage and repair before a failure or injury occurs. The hydraulic reservoir should be checked during routine lubrication service. If fluid must be added frequently, the leak should be located and fixed as soon as possible. The lift gate electrical components should be examined for signs of corrosion and wear. Check the latches and safety chains for signs of weakness.

Task D6.9   Inspect rear (ICC/DOT) impact guard.

On the rear of straight trucks and semi-trailers there are rubber bumpers that are used as cushions when backing into a loading dock. Vehicles manufactured after 1998 are required to be equipped with collapsible docking guards. These should be examined for damage during a PMI inspection since they are subject to constant abuse.

E. Road/Operational Test (3 questions)

Task E1   Check operation of clutch and gear shift.

Begin checking clutch operation by starting the engine and depressing the clutch pedal. Listen for abnormal noise. If there is noise when the pedal is depressed and it disappears when the pedal is released there may be a noisy release bearing or wear associated with the clutch intermediate plate. Attempt to put the transmission in gear. You should be able to do this without gear clash. Check the operation of the clutch brake or countershaft brake if gear clash occurs. With the transmission in gear and the brakes released slowly engage the clutch. The truck should begin moving smoothly. Observe any chatter, abnormal vibrations, or noise. Upshift and downshift in the normal manner through the gears and note any abnormal conditions for later diagnosis and repair. A pyrometer is used to indicate exhaust gas temperature and can provide the driver with engine loading information useful for shifting.

Task E2   Check operation of all instruments, gauges, and lights.

Check that all instruments, gauges, and indicator lights function properly and indicate a reading on the instrument panel. Record on the road test report.

Task E3   Check steering wheel for play, binding and centering.

Test the operation of the steering wheel at both low and high speeds. Be sure the steering wheel is centered when driving the truck straight ahead. Note abnormalities.

Task E4   **Check operation of automatic transmission.**

During the drive test, note the operation of the automatic transmission. Upshifts should occur without over-revving the engine and engine rpms should not over-run during a shift. The engine also should not lug down following a shift. Both upshifts and downshifts should be firm and should occur without harshness or shock to the drive train. If the truck is equipped with an automatic transmission temperature gauge, you should note its reading when the temperature has stabilized and compare it to factory specifications for your application.

Task E5   **Check road speed limiter.**

Note the operation of the road speed governor. Reprogram if required.

Task E6   **Check cruise control.**

During the road test, check hard cruise, soft cruise, and upper and lower droop. Reprogram if required.

Task E7   **Observe exhaust for excessive smoke.**

The color and consistency of the exhaust can reveal combustion problems. When a diesel engine is running properly with a heavy load, exhaust emission should be clear. Note and report any smoking during the road test. It can be an indication of serious problems that could be avoided if addressed at a PMI.

Task E8   **Test service brakes.**

The brakes that are activated by the brake pedal are referred to as the service brakes. Brake tests, either hydraulic or air type, have stopping distance criteria that must be attained. The FMVSS sets the standards used in the United States and Canada to define stopping distances as well as other braking safety criteria. Work the brakes during the road test using a mixture of snub braking and aggressive panic stops.

Task E9   **Verify engine/exhaust brake or retarder operation.**

The periodic servicing and/or inspection procedures for exhaust brakes are simple to perform.
Use the engine brake during the road test when the engine rpm is in torque rise. Note the disengage rpm. In multicylinder, internal compression brake systems verify engine brake operation in each cylinder and banks of cylinders. Engine brakes are not a deadline item but are likely to produce driver complaints if not functioning properly.

Trucks equipped with a retarder offer the same brake-saving advantages of an engine/exhaust brake, but the technology is different. Most retarders are incorporated into an automatic transmission. Caterpillar truck engines can be equipped with a brake saver mounted in the flywheel housing, which may be coupled with manual transmissions. Retarders may be operated either manually, or automatically. During the drive test, check the retarder to ensure that it slows the truck down when applied. The sliding gate exhaust brake uses a pneumatically activated gate actuated by chassis system pressure. The air supply to close the gate is controlled by an electronically switched pilot valve. An aperture in the sliding gate permits minimal flow through the brake gate during engine braking. The butterfly valve version operates similarly.

Task E10   **Check operation of backup warning.**

The backup warning indicator circuit should energize when the vehicle is placed into the reverse gear. An audible-pulse warning sound should be heard and the backup light illuminate. Backup alarms must function and are a deadline item in many jurisdictions.

Sample Test

Please note the letter and number in parentheses following each question. They match the task in Section 4 that discusses the relevant subject matter. You may want to refer to the overview using the cross-referencing key to help with questions posing problems for you.

1. Cracks between wheel lug holes indicate
   A. overinflation of tires.
   B. overloading.
   C. using larger than recommended tire sizes.
   D. incorrect wheel size for application.   (D5.8)

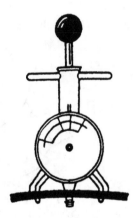

2. What is shown in the figure above?
   A. cooling system tester
   B. belt tension gauge
   C. vacuum gauge
   D. dial indicator   (A1.3)

3. The gauge in the figure above shows
   A. water temperature.
   B. oil temperature.
   C. oil pressure.
   D. transmission temperature.   (B1.3)

4. What activates the pump in an air-driven automatic chassis lube system?
   A. a solenoid valve
   B. a relay valve
   C. an ECM
   D. a toggle switch inside the cab   (D3.13, D4.14, D6.5)

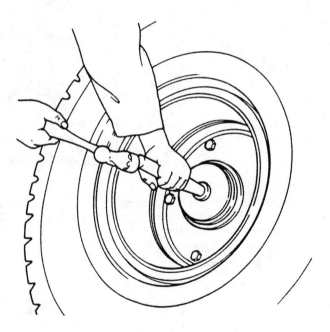

5. The operation being performed in the figure above is
   A. checking wheel-hub runout.
   B. aligning wheel hub.
   C. installing wheel bearings.
   D. installing a wheel seal.   (D4.7, D4.9)

6. The fifth wheel to upper coupler connection should have no more than
   A. ¹⁄₁₆ inches longitudinal play.
   B. ⅛ inches longitudinal play.
   C. ¼ inches longitudinal play.
   D. ½ inches longitudinal play.   (D6.2)

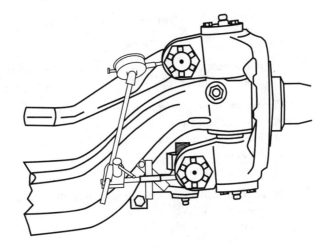

7. The procedure being performed in the figure above is
   A. knuckle vertical play.
   B. knuckle upper bushing play measurement.
   C. lower bushing deflection.
   D. kingpin axial play.

(D4.6)

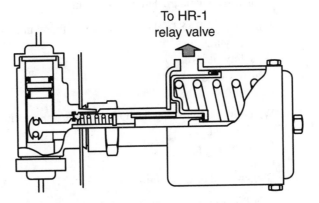

To HR-1
relay valve

8. What is indicated in the figure above?
   A. hydraulic liftgate cylinder
   B. bypass solenoid for reefer unit
   C. remote automatic kingpin locking relay
   D. hydraulic spring parking brake actuator

(D2.3)

9. If the fuel is contaminated with water, the water will
   A. remain mixed with the fuel.
   B. float on top of the fuel.
   C. settle to the bottom of the fuel tank.
   D. flow through the filters into the injection system.

(A2.3)

10. Technician A says a loose grab handle may cause personal injury. Technician B says the steps must be level and free of debris. Who is correct?
   A. A only
   B. B only
   C. Both A and B
   D. Neither A nor B

(B3.4)

11. When testing the maximum output of an alternator, the engine should be run at
    A. 1,000 rpm.
    B. 1,500–2,000 rpm.
    C. idle speed.
    D. 2,000 alternator rpm.   (C2.2)

12. A binding steering wheel may be caused by
    A. loose tie-rod ends.
    B. loose or worn steering shaft U-joints.
    C. loose drag-link ends.
    D. worn kingpin bushings.   (E3)

13. What effect will an increase in oil temperature have on the transmission oil level?
    A. lower the oil level
    B. raise the oil level
    C. no effect
    D. aerate the oil   (E4)

14. A wheel cylinder
    A. does not need any maintenance.
    B. needs to be inspected at a PM.
    C. needs to be replaced periodically.
    D. is usually constructed of aluminum.   (D2.5)

15. What is the LEAST-Likely article to be found in a safety kit?
    A. safety triangles
    B. safety flares
    C. fire extinguisher
    D. tool box   (B2.2)

16. Which of the following is NOT part of an automatic chassis lube system?
    A. grease lines
    B. reservoir
    C. pump
    D. manifold or distribution block   (D4.14, D6.5)

17. All of these problems may cause a low, spongy brake pedal EXCEPT
    A. excessive brake pedal free play.
    B. air in the hydraulic brake system.
    C. brake fluid contaminated with moisture.
    D. low brake fluid level.   (D2.4)

18. What would be the LEAST-Likely cause of a front wheel shimmy?
    A. kingpin play
    B. improperly adjusted steering gear relief plungers
    C. loose Pitman arm
    D. tire imbalance   (D4.1, D4.5, D4.6, D4.7)

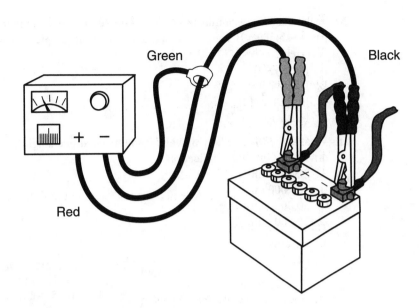

19. The test in the figure above is designed to measure
    A. alternator amperage output.
    B. battery voltage and amperage.
    C. voltage drop of the alternator field circuit.
    D. voltage drop of the battery negative circuit.   (C1.4, C2.3)

20. The bearings in the wheel hub shown in the following figure
    A. must be lubricated with wheel bearing grease before assembly.
    B. are permanently lubricated with the specified oil.
    C. require a specific hub end play.
    D. require inner and outer hub seals.   (D4.8)

21. Technician A says the back-up warning system provides a warning to anyone standing behind the
    truck. Technician B says the back-up warning system provides a visual warning. Who is correct?
    A. A only
    B. B only
    C. Both A and B
    D. Neither A nor B   (E10)

22. Technician A says air seat spring suspensions eliminate the need for cab shock absorbers.
    Technician B says you can adjust the leveling valve to maintain the proper cab height. Who
    is correct?
    A. A only
    B. B only
    C. Both A and B
    D. Neither A nor B   (B3.9)

23. Technician A says air seat spring suspensions eliminate the need for cab shock absorbers.
    Technician B says you can adjust the leveling valve to maintain the proper seat height. Who
    is correct?
    A. A only
    B. B only
    C. Both A and B
    D. Neither A nor B   (B3.2)

24. Technician A says a transmission oil sample can be taken after the oil is drained into a drain pan. Technician B says to take a transmission oil sample mid-way through the drain-off. Who is correct?
    A. A only
    B. B only
    C. Both A and B
    D. Neither A nor B

(D3.18)

25. What is the Most-Likely mileage recommended in changing lubricant for truck drive axles driven in a linehaul application?
    A. 200,000 miles
    B. 100,000 miles
    C. 50,000 miles
    D. 25,000 miles

(D3.15)

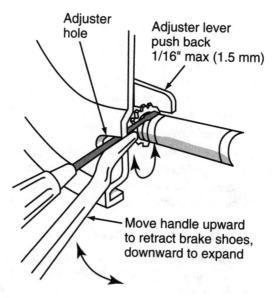

26. On the hydraulic drum brake in the figure above, the technician is
    A. performing a hydraulic drum brake adjustment.
    B. changing a return spring on a hydraulic brake.
    C. measuring brake shoe to drum clearance.
    D. setting the park brake gap.

(D2.9)

27. Technician A says moisture and corrosion on battery tops cause faster battery self-discharge. Technician B says excessive corrosion on the battery box and cover indicates undercharging. Who is correct?
    A. A only
    B. B only
    C. Both A and B
    D. Neither A nor B

(C1.1)

28. Cloudiness in hydraulic brake fluid is a sign of
    A. air in the fluid.
    B. dirt contamination.
    C. overheated/burnt fluid.
    D. cold fluid.

(D2.1)

29. Filters installed in the HVAC air delivery system are
    A. always made of fiberglass mesh to resist corrosion.
    B. designed to remove moisture from cab and sleeper air.
    C. not individually replaceable.
    D. designed to remove dust and dirt from cab and sleeper air.   (B4.4)

30. What is not checked when inspecting an air-brake system?
    A. tractor-protection valve
    B. spring-brake inversion system
    C. desiccant pack in air dryer
    D. operation of the clutch brake   (D1.3, D1.9, D1.10, E8)

31. Excessive black smoke in diesel exhaust may be caused by
    A. a restricted intake.
    B. misfiring injectors that do not inject fuel.
    C. coolant entering the combustion chambers.
    D. oil entering the combustion chamber.   (E7)

32. Technician A says that the Department of Transportation (DOT) does not require PM records
    on lubrication. Technician B says that maintenance records are an invaluable source of
    information in the life of a heavy truck. Who is correct?
    A. A only
    B. B only
    C. Both A and B
    D. Neither A nor B   (B3.7)

33. Technician A says that an inspection of the road speed governor is required during the
    road/operational test. Technician B says that more emphasis should be placed on the post-trip
    inspection rather than the pre-trip inspection. Who is correct?
    A. A only
    B. B only
    C. Both A and B
    D. Neither A nor B   (E5)

34. Technician A says that clutch-pedal free travel should be between 3 and 4 inches. Technician B
    says that the last inch of clutch-pedal travel activates the clutch brake on trucks equipped with
    a clutch brake. Who is correct?
    A. A only
    B. B only
    C. Both A and B
    D. Neither A nor B   (B3.11)

35. The drum wear shown in the figure above indicates
    A. a concave drum.
    B. a convex drum.
    C. an offset drum.
    D. a shifted drum.   (D1.16)

36. An audible air leak is observed that appears to be coming from the front brake service brake chamber with the brakes applied. Which of the following is the LEAST-Likely cause?
    A. loose mounting bolts
    B. ruptured diaphragm
    C. loose chamber clamp
    D. missing clamp bolt

    (D1.5)

37. A check of the general condition of the alternator is being performed. Technician A says to check for out-of-round mounting holes. Technician B checks the belts before making performance tests. Who is correct?
    A. A only
    B. B only
    C. Both A and B
    D. Neither A nor B

    (C2.1)

38. In winterizing a vehicle, Technician A will use a Teflon lubricant on door latches and dry graphite on the door locks. Technician B mounts alcohol evaporator downstream from the air dryer. Who is correct?
    A. A only
    B. B only
    C. Both A and B
    D. Neither A nor B

    (B3.8)

39. Technician A says you turn the key on the ignition to perform electronic system self-checks. Technician B says the ignition circuit powers most computer controlled chassis systems. Who is correct?
    A. A only
    B. B only
    C. Both A and B
    D. Neither A nor B

    (B1.1, B1.7)

40. During a PMI, Technician A says the air passages through the air-to-air intercooler should be checked for restrictions. Technician B says the air-to-air intercooler uses refrigerant to lower the intake air temperature. Who is correct?
    A. A only
    B. B only
    C. Both A and B
    D. Neither A nor B

    (A3.3)

41. Technician A says that ⅜ inch is the maximum allowable when checking fan clutch play. Technician B lubricates all door latches and door locks when an inspection is performed. Who is correct?
    A. A only
    B. B only
    C. Both A and B
    D. Neither A nor B

    (A4.1, B3.8, B3.10)

42. Technician A says overfilling a manual transmission could result in overheating of the transmission. Technician B says overfilling a transmission could cause excessive aeration of the transmission fluid. Who is correct?
    A. A only
    B. B only
    C. Both A and B
    D. Neither A nor B

    (D3.17)

43. Technician A says that it is OK to use ether on glow-plug equipped engines. Technician B always uses ether when starting a diesel engine. Who is correct?
    A. A only
    B. B only
    C. Both A and B
    D. Neither A nor B   (A1.1)

44. Technician A says to inflate mounted tires in a safety cage or using a portable lock ring guard. Technician B says to first mount the tire on the truck and then inflate to the proper tire pressure. Who is correct?
    A. A only
    B. B only
    C. Both A and B
    D. Neither A nor B   (D5.3)

45. Technician A says that when performing a brake overhaul on a hydraulic rear drum brake system, you replace the return springs. Technician B says the maximum allowable out-of-round specification on a brake drum should be 0.025 inch. Who is correct?
    A. A only
    B. B only
    C. Both A and B
    D. Neither A nor B   (D2.8)

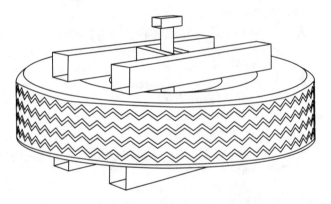

46. Technician A says the device in the figure above is to restrain the tire while it is being inflated. Technician B says that the device could be used to center the tire for reuse. Who is correct?
    A. A only
    B. B only
    C. Both A and B
    D. Neither A nor B   (D5.5)

47. While discussing water pumps, Technician A says a defective water pump bearing may cause a growling noise with the engine idling. Technician B says the water pump bearing may be ruined by coolant leaking past the water pump seal. Who is correct?
    A. A only
    B. B only
    C. Both A and B
    D. Neither A nor B   (A4.10)

48. Technician A says that a circle inspection includes items such as torn mud flaps, leaking hoses, and outdated permits. Technician B says to perform a circle inspection first before any other inspection as part of a PMI. Who is correct?
    A. A only
    B. B only
    C. Both A and B
    D. Neither A nor B   (D6.3)

49. When adjusting the height control valve in an air suspension system
    A. the height control valve arm may be moved upward to lower the suspension height.
    B. steel locating pins are installed in the height control valve arms.
    C. the air brake system pressure must be 70 psi (483 kPa) or more.
    D. the truck must have an average load on the chassis.
    (D4.13)

50. During a tire inspection, a technician notices a cut in the sidewall of a tire. Which of the following is the Most-Likely cause?
    A. excessive toe-in
    B. camber
    C. careless driving
    D. caster
    (D5.2)

51. The LEAST-Likely cause of water droplets or gray oil on the dipstick would be
    A. cracked block or cylinder head.
    B. blown head gasket.
    C. leaky oil cooler.
    D. worn oil pump.
    (A1.6)

52. Refrigerant leaks at condenser line fittings are indicated by
    A. an oily film around the fitting.
    B. a clear fluid dripping from the fitting.
    C. a hissing sound coming from the fitting.
    D. corrosion on the fitting.
    (B4.2)

53. A gasoline engine experiences rough idle operation. This problem is worse when the engine is at normal operating temperature. The cylinder compression is within specifications and all the ignition components are satisfactory. The cause of this problem could be
    A. an EGR valve stuck open.
    B. a burned exhaust valve.
    C. valve adjustment that is too tight.
    D. a bent intake valve.
    (A2.6)

54. Which of the following could be the cause of an engine coolant temperature gauge reading high continuously?
    A. open sending unit circuit
    B. short to ground in the instrument panel power feed circuit
    C. short circuit in the sending unit
    D. open or burned out gauge
    (B1.2, B1.3)

55. Interference of a shock absorber outer tube to the reservoir can cause
    A. premature bushing wear.
    B. scraping noise with travel.
    C. main shaft seal failure.
    D. excessive fatigue in mount assembly.
    (D4.11)

56. Technician A says to test a steering linkage joint for excessive wear, you should compress the joint with a C-clamp or large pliers. Technician B says simply pushing against the joint with force that you can create with your hands should be enough to identify excessive wear. Who is correct?
    A. A only
    B. B only
    C. Both A and B
    D. Neither A nor B
    (D4.5)

57. Technician A says that when filling the cooling system the air bleeder valves should be closed. Technician B says that if the coolant solution contains over 67 percent antifreeze, heat transfer from the cylinders to the coolant is adversely affected. Who is correct?
    A. A only
    B. B only
    C. Both A and B
    D. Neither A nor B   (A4.9)

58. While checking fan blades, Technician A says a cracked fan blade may cause a clicking or creaking noise with the engine idling. Technician B says you should not stand or position your body in line with the fan while increasing the engine speed. Who is correct?
    A. A only
    B. B only
    C. Both A and B
    D. Neither A nor B   (A4.3)

59. The seat belts in the cab and sleeper are being inspected. Technician A says with the seat belt on and the truck moving, activate the parking brake to stop the truck. Technician B says it is not necessary to check the seat belts or restraints in the sleeper, only the cab. Who is correct?
    A. A only
    B. B only
    C. Both A and B
    D. Neither A nor B   (B2.3)

60. Technician A says if a truck is equipped with automatic temperature control (ATC), the outlet temperature should be within several degrees of the set temperature. Technician B says that the blower motor should operate in all speed positions. Who is correct?
    A. A only
    B. B only
    C. Both A and B
    D. Neither A nor B   (B1.5)

61. A driver complains that the power steering has become increasingly harder to turn over the last 1,000 miles. Technician A says you may need to replace the system filter because of clogging. Technician B says the problem may be a loose steering gear mount that could be binding the steering column. Who is correct?
    A. A only
    B. B only
    C. Both A and B
    D. Neither A nor B   (D4.4)

62. What circuit is being operated when a starter draw test is being performed?
    A. charging circuit
    B. cranking circuit
    C. regulator circuit
    D. control circuit   (C1.5, C1.6)

63. All of the following are cause for an OOS deadline EXCEPT
    A. inoperative tractor protection valve.
    B. radial tires with ½ inches minimum tread depth measured in adjacent grooves.
    C. movement under steering load of a stud nut.
    D. air reservoir mounting brackets that are loose.   (D1.6, D1.10, D5.4)

64. Technician A says that the hydraulic assist back-up system and warning devices should be checked during each inspection. Technician B says that checking the hydraulic assist back-up system at each inspection is unnecessary as the warning devices will let the driver know if a problem exists. Who is correct?
   A. A only
   B. B only
   C. Both A and B
   D. Neither A nor D
                                                                                      (D2.10)

65. When inspecting an oil filter housing or its mounting, which of the following is LEAST-Likely to be a procedure?
   A. visually checking for cracks
   B. visually checking gasket surface for nicks
   C. magna fluxing the housing and metal parts to detect small cracks
   D. visually inspecting housing passageways for obstruction and cracks
                                                                                      (A5.1)

66. When inspecting a tractor, a technician notes that the window does not go up and down freely and the handle is loose. What should the technician do?
   A. Service it immediately.
   B. Record the malfunctioning window on the PMI, then service it immediately.
   C. Inform the service department that the window needs servicing.
   D. Record the observed problem on a PMI.
                                                                                      (B3.3)

67. The S-cam on one foundation brake fitted with Q brakes rolls over. Which of the following is the Most-Likely cause?
   A. excessively worn drums and shoes
   B. application air pressure too high
   C. improperly installed slack adjuster
   D. brake chamber push rod too long
                                                                                      (D1.18)

68. Technician A says that a circle inspection includes items such as torn mud flaps, leaking hoses, and expired permits. Technician B says to perform a circle inspection before any other inspection as part of a PMI. Who is correct?
   A. A only
   B. B only
   C. Both A and B
   D. Neither A nor B
                                                         (D5.1, D5.2, D5.6, D5.8, D6.3, D6.9)

69. A turbocharger on a tractor has less than the specified boost pressure. All of these defects may be the cause of the problem EXCEPT
   A. turbocharger bearings that are starting to fail.
   B. improper wastegate adjustment.
   C. damaged compressor wheel.
   D. turbocharger oil seals leaking oil into the exhaust.
                                                                                      (A3.4)

70. While discussing van and over-top trailer bodies, Technician A says if one floor cross member is broken and sagging below the lower rail, the vehicle should be tagged out of service. Technician B says if a drop frame is visibly twisted, the vehicle should be placed out of service. Who is correct?
   A. A only
   B. B only
   C. Both A and B
   D. Neither A nor B
                                                                                      (D6.7)

71. While inspecting a transmission for leaks, the technician notices that the gaskets appear to be blown out of their mating sealing surfaces. What should the technician check first?
    A. the transmission breather
    B. the shifter cover
    C. the release bearing
    D. the rear seal   (D3.7)

72. An exhaust system on a cab-over is being inspected. Technician A says to inspect the system while it is hot, so proper pipe to body/frame clearances may be measured. Technician B uses a prybar to check the looseness of the piping and the strength of the brackets when performing an exhaust system PMI. Who is correct?
    A. A only
    B. B only
    C. Both A and B
    D. Neither A nor B   (A3.1)

73. A scheduled inspection is being performed. Technician A says that if lift gates and controls are not on the checklist, then it is not mandatory to inspect them. Technician B says you always check cargo ramps on a vehicle equipped with them. Who is correct?
    A. A only
    B. B only
    C. Both A and B
    D. Neither A nor B   (D6.8)

74. A gauge cluster is being inspected for proper function and illumination. Technician A says that a malfunctioning oil pressure gauge is not a valid reason for tagging a vehicle out of service. Technician B says that a proper function check of the instrumentation panel is required under a PMI. Who is correct?
    A. A only
    B. B only
    C. Both A and B
    D. Neither A nor B   (E2)

75. Technician A says a 45-degree elbow in an air brake line may be replaced with a 90-degree elbow without affecting brake operation. Technician B says that if a trailer service system has an air fitting that is smaller than the original line, the air timing/balance is affected. Who is correct?
    A. A only
    B. B only
    C. Both A and B
    D. Neither A nor B   (D1.13)

76. When inspecting the brake pads on a truck equipped with hydraulic disc brakes, Technician A says you must remove the wheel in most cases to measure pad thickness. Technician B says that if the brakes were applied shortly before measuring the lining-to-rotor clearance, this clearance will be greater than specified. Who is correct?
    A. A only
    B. B only
    C. Both A and B
    D. Neither A nor B   (D2.7)

77. An excessively high coolant level in the recovery reservoir may be caused by any of these problems EXCEPT
    A. restricted radiator tubes.
    B. a thermostat that is stuck open.
    C. a blown head gasket.
    D. an inoperative air-operated cooling fan clutch.   (A4.6)

78. A small leak is found in an A/C system. What should the technician do?
    A. Nothing. Wait for a service report to be filed.
    B. Log the problem on a PMI schedule sheet.
    C. Log the problem, then inform the owner/driver that this is in violation of the Clean Air Act.
    D. Repair the leak immediately.
    (B4.3)

79. When inspecting a tractor for physical damage, a technician notes that the front bumper is cracked near the mounting bolts and in danger of breaking off. What should the technician do?
    A. Record the observed problems on a PMI.
    B. Inform the service department that the bumper needs servicing.
    C. Immediately repair the cracked bumper.
    D. Service it after a thorough inspection of the entire vehicle is performed.
    (B3.13)

80. A technician is testing the low air pressure warning system on a truck by performing a pressure drop test on a straight single truck's air brake system. Technician A says that the pressure drop should be no more than 6 psi in 2 minutes. Technician B says you hold down the brake pedal for the first minute then release it. Who is correct?
    A. A only
    B. B only
    C. Both A and B
    D. Neither A nor B
    (D1.7)

81. Technician A says that the air cleaner element should be replaced when 25 in. Hg. air restriction is indicated. Technician B says that regardless of regulations, a truck must be serviced in the most cost-effective manner to the owner. Who is correct?
    A. A only
    B. B only
    C. Both A and B
    D. Neither A nor B
    (A3.5)

82. The tractor protection valve closes to protect the tractor air supply if the air pressure in the supply line drops below
    A. 20 psi (138 kPa).
    B. 55 psi (379 kPa).
    C. 75 psi (517 kPa).
    D. 80 psi (552 kPa).
    (D1.9)

83. A cooling system is being pressure tested. Technician A says that a radiator pressure cap reading of 15 psi is normal. Technician B says that it is necessary to warm the engine to normal operating temperature before conducting a pressure test. Who is correct?
    A. A only
    B. B only
    C. Both A and B
    D. Neither A nor B
    (A4.4)

84. When testing the operation of the trailer control valve, a technician should
    A. move the handle to the fully applied position and record the air pressure.
    B. drain the air system to zero psi, and then allow pressure to build up again.
    C. record the gauge reading of the air pressure first, and then drain the system.
    D. listen for air leakage around the handle as the handle is moved to the applied position.
    (D1.11)

85. Technician A says any type of wire may be used to replace a burned fusible link as long as the gauge is one size smaller than the wire in the circuit being protected. Technician B says that a circuit breaker should be replaced after repairing a short in a circuit. Who is correct?
    A. A only
    B. B only
    C. Both A and B
    D. Neither A nor B
    (A1.7)

86. While performing routine maintenance, a technician observes that the power steering system has a filter. Technician A says the filter should be replaced, if required, on a PMI according to the OEM. Technician B says the filter should be replaced after a component failure. Who is correct?
   A. A only
   B. B only
   C. Both A and B
   D. Neither A nor B   (D4.3)

87. Most vehicle manufacturers recommend checking the level of power steering fluid with the system at a working temperature of
   A. 100°F.
   B. 140°F.
   C. 90°F.
   D. 175°F.   (D4.2)

88. The vehicle to which a pintle hook or drawbar is attached should be tagged out of service for any of the these defects EXCEPT
   A. wear in the pintle hook horn that exceeds 10 percent of the horn section.
   B. loose or missing mounting bolts.
   C. any welded repairs to the drawbar eye.
   D. cracks in the pintle hook mounting surface.   (D6.4)

89. An electronic cruise control maintains the vehicle speed at 45 mph (72 km/h) when the selected speed is 60 mph (96 km/h). The Most-Likely cause of this problem is
   A. a defective cruise control module.
   B. a defective vehicle speed sensor (VSS).
   C. improper size of tires on the drive axles.
   D. a defective cruise control switch.   (E6)

90. A combination vehicle consisting of a tractor, trailer, and a converter dolly is being inspected. Technician A says that the tractor must carry a copy of each inspection form in the cab. Technician B says that one inspection form is good for all trailers, dollies, and towed vehicles. Who is correct?
   A. A only
   B. B only
   C. Both A and B
   D. Neither A nor B   (B2.6)

91. When discussing the ABS warning light in the instrument panel, Technician A says that if the ABS warning light is on longer than 4 seconds after the engine is started, there is a defect in the ABS. Technician B says that if the ABS warning light is on at vehicle speeds above 4 mph (6.43 km/h) there is a defect in the ABS. Who is correct?
   A. A only
   B. B only
   C. Both A and B
   D. Neither A nor B   (D1.12)

92. What does the arrow in the figure above indicate?
    A. clearance light
    B. identification light
    C. fog light
    D. intermediate light   (C3.1, C3.2)

93. What type of lubricant should be used to grease the fifth wheel?
    A. 80/90 gear lube
    B. water-resistant lithium-base grease
    C. any type of bearing grease
    D. 85/140 axle lube   (D6.5)

94. Fluid leaks are evident between the axle housing and the carrier assembly. What would the Most-Likely cause of this leak be?
    A. damaged gasket or missing sealant
    B. repeated overloading of the drive train
    C. plugged axle housing breather vent
    D. moisture contaminated axle lubricant   (D3.11)

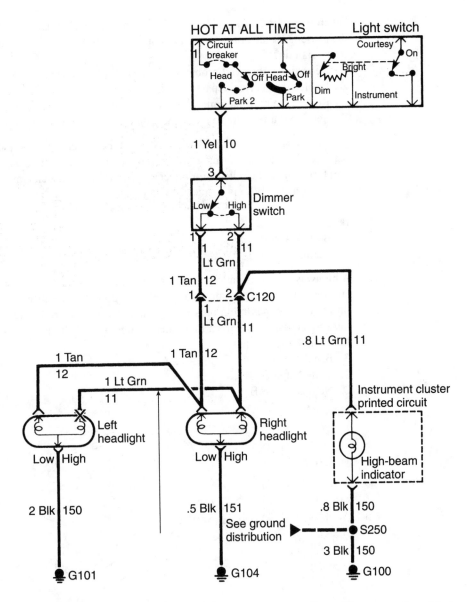

95. What would occur if the circuit, indicated at point X in the figure above, was found to be heavily corroded?
    A. The right headlight would be inoperative.
    B. The left headlight would be inoperative.
    C. The right high beam would be dim.
    D. The left high beam would be dim.   (C3.2)

96. A technician notices a whitish milky substance when changing the fluid in an axle. This evidence of water is likely caused by
    A. common condensation.
    B. infrequent driving and short trips.
    C. the axle being submerged in water.
    D. the vehicle being frequently driven during rainy or wet conditions.   (D3.19)

97. What is the LEAST-Likely cause of a vehicle wandering?
    A. mismatched or under-inflated tires
    B. unmatched spring design/load capacity spring assemblies
    C. one or more broken spring leaves
    D. frame alignment within manufacturer's specifications   (D4.12, D5.5, D5.9, E3)

98. When inspecting a tractor for physical damage, a technician notes that the mirror mounts for the passenger side are loose and in danger of separating from the door. What should the technician do?
    A. Record the observed problem on a PMI.
    B. Inform the service department that the bracket needs servicing.
    C. Record the malfunctioning bracket on a PMI, then service it immediately.
    D. Service it after a thorough inspection of the entire vehicle is performed.   (B3.5)

99. Technician A says a correctly adjusted pull type clutch requires ½ inch (1.25cm) of clearance between the release bearing and the clutch brake disc. Technician B says a correctly adjusted pull type clutch the release fork needs ⅛ inch (0.3125cm) clearance between itself and the release bearing. Who is correct?
    A. A only
    B. B only
    C. Both A and B
    D. Neither A nor B   (D3.4))

100. A highway tractor is having a full PMI performed. Technician A says that outboard S-cam bushings should be pressure-lubricated until grease flows from the outboard seal. Technician B says to use the maximum pressure available from pressure lubricating equipment to aid in flushing contaminants from lubricated components. Who is correct?
    A. A only
    B. B only
    C. Both A and B
    D. Neither A nor B   (D1.19)

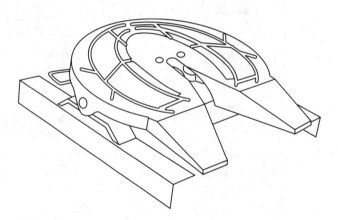

101. Which of these would not be part of performing a PMI, on a truck with a fifth wheel like the one in the figure above?
    A. Reverse the location of the jaws every 100,000 miles.
    B. Make sure horizontal movement does not exceed ⅛ inch.
    C. Torque mounting bolts.
    D. Inspect the locking mechanism.   (D6.1, D6.2)

102. When performing a test drive on a vehicle, Technician A says to test the service brakes by applying and releasing them. Technician B says that although a test drive is not mandatory, it is a good idea to perform one after a vehicle has been serviced. Who is correct?
    A. A only
    B. B only
    C. Both A and B
    D. Neither A nor B   (E8)

103. All of these procedures should be done when replacing a torque rod on a tandem axle suspension EXCEPT
    A. replace alignment washers with new ones.
    B. secure the truck's parking brakes and chock the tires.
    C. ensure the replacement torque rod is of the correct length.
    D. check the axle alignment when finished.   (D4.15)

104. In electronic fuel-injected diesel and gasoline engines, which sensor is common to both systems?
    A. knock sensor
    B. O_2 sensor
    C. mass air flow sensor
    D. throttle position sensor   (A2.2, A2.6)

105. Technician A says idle speed can usually be raised and lowered by the use of a toggle switch provided on the dash on electronic engines. Technician B says idle shutdown duration is programmable on electronic engines. Who is correct?
    A. A only
    B. B only
    C. Both A and B
    D. Neither A nor B   (B1.4)

106. A roadside inspection is performed while it is raining. All of the following defects would be reason to tag the vehicle out of service EXCEPT
    A. the wiper blade on the driver's side is cracked and badly worn.
    B. the wiper motor is inoperative.
    C. the passenger's side wiper blade is cracked and badly worn.
    D. the driver's side wiper arm is distorted so the blade does not contact the windshield properly.   (2.4)

107. A heavy-duty truck is found to have an exhaust leak. Technician A intends to either repair it immediately, or deadline the truck. Technician B says that the vehicle can only be put out of service if the leaking exhaust fumes enter the driver/sleeper compartment or if they are likely to result in charring, burning, or damaging the wiring, fuel supply, or other combustible parts. Who is correct?
    A. A only
    B. B only
    C. Both A and B
    D. Neither A nor B   (A3.2)

108. How should a truck's engine mounts be dynamically checked?
    A. visually inspecting for cracks and/or deterioration
    B. correlating information from the driver and the vehicle's maintenance history
    C. observing the engine mounts while the engine is torqued with its brakes applied
    D. automatically deadlining a vehicle due to the driver's suspicions   (A1.5)

109. Technician A says that an air shift unit with a sticking shift shaft can be cleaned and reinstalled. Technician B says that the air shift unit is not serviceable and should be replaced. Who is correct?
    A. A only
    B. B only
    C. Both A and B
    D. Neither A nor B   (D3.16)

110. When adjusting wheel bearings, Technician A says that TMC wheel-end procedure should be followed at all times. Technician B says that wheel bearings set with preload are likely to fail prematurely. Who is correct?
   A. A only
   B. B only
   C. Both A and B
   D. Neither A nor B   (D4.9)

111. A truck mechanically actuated clutch is being measured for free play. Technician A says the measurement should be between 1 and 2 inches. Technician B says that on a hydraulic clutch free play should be 2–3 inches. Who is correct?
   A. A only
   B. B only
   C. Both A and B
   D. Neither A nor B   (D3.3, D3.4)

112. An applied pressure drop test is being performed on a straight truck air brake system. Technician A says that the pressure drop should be no more than 6 psi in 2 minutes. Technician B says you hold down the brake pedal for the first minute, and then release it. Who is correct?
   A. A only
   B. B only
   C. Both A and B
   D. Neither A nor B   (D1.4, D1.5)

113. When checking the emission control systems of a gasoline fueled truck, always test
   A. the carburetor fuel inlet line.
   B. the PCV valve.
   C. the air cleaner vacuum hoses.
   D. the intake manifold service port.   (A2.5)

114. When checking the coolant condition with a supplemental coolant additive (SCA) test strip, the technician finds that the coolant pH condition is higher than specification limits. While reviewing maintenance records the technician finds that the pH condition was higher than the limit at the last service also. Which of these items should the technician do?
   A. Add more antifreeze to increase SCA.
   B. Continue to run the truck until the next PMI.
   C. Drain the entire cooling system and add the proper SCA mixture.
   D. Run the truck with no SCA additives until the next PMI.   (A4.8)

115. Technician A says that when checking the condition of hydraulic drum brakes on a vehicle, to look for leakage and contamination around the wheel cylinders. Technician B inspects the brake lines for evidence of corrosion. Who is correct?
   A. A only
   B. B only
   C. Both A and B
   D. Neither A nor B   (D2.2, D2.5)

116. An exhaust brake is being inspected on a truck. Technician A says that this should be performed during the road test. Technician B says that a properly functioning engine compression brake should be noiseless. Who is correct?
   A. A only
   B. B only
   C. Both A and B
   D. Neither A nor B   (E9)

117. Which of the following would be the least important part of a trailer's lighting system to check?
    A. side amber marker light(s)
    B. interior overhead light(s)
    C. main wiring input socket and associated wiring
    D. reefer power supply   (C3.3)

118. All of the following practices should be avoided when working with truck frames EXCEPT
    A. drilling holes in the frame web.
    B. drilling holes in the frame flanges.
    C. cutting holes in the frame with a torch.
    D. notching the frame rails.   (D6.6)

119. Technician A says that air horns should be checked during a PMI. Technician B says that a tire on the steering axle with $\frac{3}{32}$ inches of thread should be replaced. Who is correct?
    A. A only
    B. B only
    C. Both A and B
    D. Neither A nor B   (B2.1, D5.4)

120. A vehicle with vacuum boosted brakes is tagged out of service if in the vacuum system the
    A. vacuum never exceeds 15 in. Hg.
    B. vacuum lines appear to be OK, but the engine lacks performance.
    C. vacuum reserve is insufficient to permit one full brake application after the engine is shut off.
    D. system vacuum bleeds down over 50 percent after engine shut off within 2 minutes.   (D2.6)

121. In relation to lubricants (transmission and drive axle), what is channeling?
    A. lubricant breakdown in high temperatures
    B. extreme pressures on components causes a lubricant to separate
    C. liquid seeking the lowest center of gravity
    D. components starved of lubrication due to cold temperatures   (D3.9, D3.14)

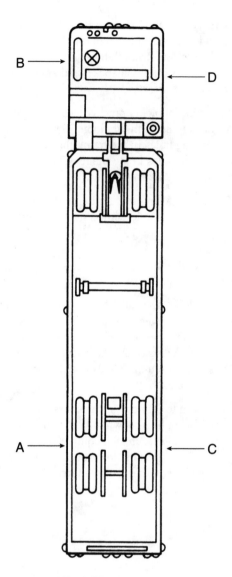

122. In the figure above, what would be the proper sequence of performing a circle-check on this combination?
    A. D-C-A-B
    B. B-A-C-D
    C. B-D-C-A
    D. C-A-B-D   (B3.7)

123. Technician A says an engine oil sample should be taken during every PMI. Technician B says to take an engine oil sample mid-way through the sump drain-off. Who is correct?
    A. A only
    B. B only
    C. Both A and B
    D. Neither A nor B   (A5.2)

124. If a starter ground circuit test reveals a voltage drop of more than 0.5 volts, the problem may be
    A. a shorted armature winding.
    B. an open field coil.
    C. a damaged battery ground cable.
    D. a faulty ignition switch.   (C1.2)

125. Which of the following should be done when performing a PMI on a heavy truck fuel system?
  A. Check the pour point of the fuel.
  B. Verify the accuracy of the fuel gauge.
  C. Check the fuel pump for volumetric capacity.
  D. Replace the fuel filter and prime the system.   (A2.4)

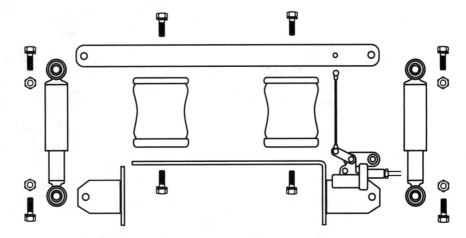

126. How would suspension height of the illustration above Most-Likely be checked?
  A. Measure from the road surface to the lower plate.
  B. Measure between the upper and lower plates.
  C. Measure from the road surface to the upper plate.
  D. Measure the angle of the regulator valve arm.   (B3.12)

127. Technician A says that lubricant leaking from the drive axles seals could be caused by a plugged breather. Technician B says that grease fittings should be wiped before lubricating. Who is correct?
  A. A only
  B. B only
  C. Both A and B
  D. Neither A nor B   (D3.12, D3.13)

128. Technician A says that technicians that perform mobile refrigerant system service must be certified. Technician B always makes sure both hand valves on the manifold gauge set are closed before connecting the hoses to the vehicle. Who is correct?
  A. A only
  B. B only
  C. Both A and B
  D. Neither A nor B   (B4.1, B4.2, B4.3)

129. Wheel assemblies are being discussed. Technician A says hub-piloted disc wheels are less susceptible to loosening and cracking than stud-piloted wheels. Technician B says that disc wheels generally experience more alignment and balance problems than do cast spoke designs. Who is correct?
  A. A only
  B. B only
  C. Both A and B
  D. Neither A nor B   (D5.6, D5.8)

130. The push hydraulic circuit in a cab lift system
  A. lowers the cab back from a fully extended position.
  B. raises the cab from fully extended to the 45 degree tilt position.
  C. allows the cab to descend to the 60 degree overhead position.
  D. raises the cab from the lowered position to the desired tilt location.   (B3.10)

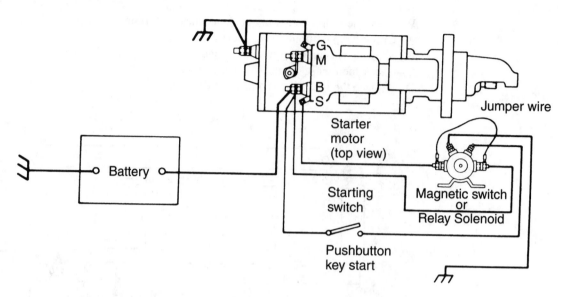

Starter motor (top view)

Jumper wire

Starting switch

Magnetic switch or Relay Solenoid

Battery

Pushbutton key start

131. What is being performed in the figure above?
   A. testing the battery output
   B. checking the continuity of the ignition switch
   C. performing a voltage drop on the ground circuit side of the starter
   D. bypassing the magnetic switch
                                                                                    (C1.6)

132. A viscous vibration damper is being inspected. What would be the most serious consequence of a visible dent in the housing?
   A. engine crankshaft failure
   B. engine misfire
   C. engine surging
   D. knocking noise
                                                                                    (A1.2)

133. Technician A says that retorquing wheel lug nuts are not required when performing a PMI. Technician B says that increased load carrying capacity is an advantage of switching to low profile tires. Who is correct?
   A. A only
   B. B only
   C. Both A and B
   D. Neither A nor B
                                                                                    (D5.7, D5.9)

134. Technician A always begins battery inspection with a general visual first, followed by an open circuit voltage test. Technician B says that it may be necessary to remove the surface charge from a battery before it is open circuit voltage tested. Who is correct?
   A. A only
   B. B only
   C. Both A and B
   D. Neither A nor B
                                                                                    (C1.3)

135. Which of the following is the best description of an acceptable axle lubricant level?
   A. Lube level is even with the bottom of the filler hole.
   B. Lube level can be easily seen through the filler hole.
   C. Lube can be touched one knuckle joint below the bottom of the filler hole.
   D. Lube can be touched with the index finger through the filler hole.
                                                                                    (D3.14)

136. A windshield is being inspected during a routine inspection. Technician A says that the vehicle need not be deadlined, since there are no cracks in the windshield greater than 2 inches in length on the driver's side of the vehicle. Technician B says that any vision distorting defect in the windshield glass must be within the sweep of the wiper on the driver's side to deadline the vehicle. Who is correct?
    A. A only
    B. B only
    C. Both A and B
    D. Neither A nor B   (B3.1)

137. Technician A says that VORAD stands for variable output relay accessory device. Technician B says that anticollision devices are now mandatory on all heavy-duty vehicles. Who is correct?
    A. A only
    B. B only
    C. Both A and B
    D. Neither A nor B   (B1.6)

138. Technician A says that a clutch brake needs to be inspected for wear and fatigue. Technician B says that providing the clutch brake drive lugs are intact, it will always function properly. Who is correct?
    A. A only
    B. B only
    C. Both A and B
    D. Neither A nor B   (D3.1)

139. Technician A says that a proper inspection of the cooling system should include a check of coolant-usage records. Technician B changes the coolant at every PMI. Who is correct?
    A. A only
    B. B only
    C. Both A and B
    D. Neither A nor B   (A1.4, A4.9)

140. Technician A says that slack adjusters should be lubricated when brakes are inspected. Technician B says that only the rear brake chamber pushrods have to be checked for stress cracks. Who is correct?
    A. A only
    B. B only
    C. Both A and B
    D. Neither A nor B   (D1.14)

141. To thoroughly inspect a transmission mount, a technician should
    A. remove the mount and tension the two mounting plates.
    B. remove the mount and inspect for signs of damage.
    C. visually inspect the mount while still in the vehicle.
    D. replace the mount if it is suspect.   (D3.8)

142. During an inspection, a tractor must be placed out of service if any of these tire conditions are present EXCEPT
    A. if a tire is mounted or inflated so it contacts any part of the vehicle.
    B. if a drive axle tire has a tread depth of less than $\frac{1}{2}$ inch at two adjacent tread grooves at three separate locations.
    C. if a steer axle tire has a tread depth of $\frac{5}{32}$ inch at two adjacent tread grooves at any location on the tire.
    D. if the tire is labeled "Not for Highway Use."   (D5.4)

143. When a self-adjusting clutch is found to be out of adjustment, check for all of the following EXCEPT
    A. correct actuator arm setting.
    B. bent adjuster arm.
    C. frozen adjusting ring.
    D. clutch shaft runout.   (D3.2 through D3.4)

144. A slack adjuster is being adjusted. Technician A says wear between the clevis pin and yoke will cause the stroke to be too long. Technician B says that an automatic slack adjuster will need frequent adjustment. Who is correct?
    A. A only
    B. B only
    C. Both A and B
    D. Neither A nor B   (D1.17)

145. While inspecting a manual transmission for leaks, the technician notices that some gaskets appear to have blown out of their mating surfaces. What should the technician check first?
    A. the transmission breather
    B. the shift housing cover
    C. the shift tower
    D. the rear seal   (D3.5, D3.7)

146. Technician A says that a possible cause of an above-normal engine temperature is a faulty fan clutch. Technician B claims that plugged radiator fins may cause overheating. Who is correct?
    A. A only
    B. B only
    C. Both A and B
    D. Neither A nor B   (A4.2)

147. The trailer lights of a double combination vehicle are being inspected. Technician A says that the wiring connector for all trailer lights is usually an 8-pin connector. Technician B says that it is mandatory to check the lights on the last trailer first in a multiple combination vehicle. Who is correct?
    A. A only
    B. B only
    C. Both A and B
    D. Neither A nor B   (B3.6)

148. Technician A says that when considering whether to deadline a vehicle, it is necessary to weigh the importance of getting a load moved versus a mechanical problem that could cause a breakdown but is not safety related. Technician B says that a vehicle should be deadlined for a cruise control malfunction. Who is correct?
    A. A only
    B. B only
    C. Both A and B
    D. Neither A nor B   (E1 through E10)

149. Technician A checks coolant hoses for cracks and hardness every time a general inspection is made. Technician B says that a general coolant system inspection should be made every time a vehicle is scheduled for a PMI. Who is correct?
    A. A only
    B. B only
    C. Both A and B
    D. Neither A nor B   (A4.5)

150. Technician A says that air drum brake linings should be replaced when there is less than ¼ inches lining when measured at the center of the shoe. Technician B says that brake linings should be replaced when they are saturated with oil or grease. Who is correct?
   A. A only
   B. B only
   C. Both A and B
   D. Neither A nor B   (D1.15)

151. A driveshaft is being inspected by a technician during a PMI. Technician A says the transmission output shaft and rear-axle input shaft do not need to be checked. Technician B always checks the U-joint trunnions for cracks, the tubes for dents, and the center bearing for damage. Who is correct?
   A. A only
   B. B only
   C. Both A and B
   D. Neither A nor B   (D3.10)

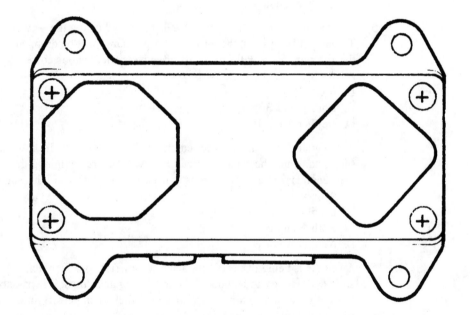

152. Technician A says that the dash control button on the left in the figure above is the trailer service brake valve. Technician B says that the dash control button on the right in the same figure is the system park valve. Who is correct?
   A. A only
   B. B only
   C. Both A and B
   D. Neither A nor B   (D1.1, D1.10)

153. When testing a suspension system on a heavy-duty truck, a technician should
   A. bounce the vehicle up and down to test the shocks.
   B. visually inspect the suspension components for signs of fatigue, deterioration, and/or failure.
   C. road test the vehicle to determine possible defects and/or misalignment(s).
   D. check the PMI schedule for scheduled repairs.   (D4.10, D4.11, D4.12)

154. Which of these criteria is the LEAST-Likely to be considered tire mismatching?
   A. radials and bias ply on the same axle
   B. dual wheels of the same nominal size but different tread design
   C. installation of a new tire next to a worn-out tire on the same axle
   D. tires with different circumferences on the same axle   (D5.9)

155. A transmission temperature gauge does not operate. Technician A says that the sensor should be replaced because he sees the same problem recur frequently. Technician B reads a value of one volt at the electrical connector for the sensor and replaces the sensor because the wiring must be intact. Who is correct?
    A. A only
    B. B only
    C. Both A and B
    D. Neither A nor B
                                                                                    (D3.6)

156. Technician A says that when refilling a cooling system it may be necessary to open a bleed valve. Technician B says that when checking a cooling system that uses ELC, to check the coolant conditioner level by using an SCA test strip specifically designed for ELC coolants. Who is correct?
    A. A only
    B. B only
    C. Both A and B
    D. Neither A nor B
                                                                              (A4.7, A4.9)

157. Technician A says to block the wheels before adjusting the service brakes on a truck. Technician B says that the air reservoir pressure must be dropped below the cut-in value before the air compressor will resume its effective cycle. Who is correct?
    A. A only
    B. B only
    C. Both A and B
    D. Neither A nor B
                                                            (D1.2, D1.5, D1.7, D1.8, D1.19)

158. An inversion valve is being tested on a heavy-duty truck. Technician A says to drain both the supply tank and the primary circuit reservoir before beginning the test. Technician B verifies that secondary circuit pressure is at least 100 psi before conducting the test. Who is correct?
    A. A only
    B. B only
    C. Both A and B
    D. Neither A nor B
                                                                                    (D1.8)

159. A vehicle is being inspected for a safety certification. Technician A states if wetness is seen around a seam of a fuel tank without forming any drops, it is passable. Technician B says that it is not necessary to drain water from the fuel when performing an inspection during the summer months. Who is correct?
    A. A only
    B. B only
    C. Both A and B
    D. Neither A nor B
                                                                                    (A2.1)

160. A truck is being road tested as part of a scheduled PMI. All of the following should be checked EXCEPT
    A. check road speed limiter.
    B. perform snap-shot test for active codes.
    C. observe exhaust for excessive smoke.
    D. verify engine brake operation.
                                                                        (B1.7, E1 through E10)

161. During a PMI, the technician tests the air supply system and finds that buildup time is slow. Which of the following is the Most-Likely cause?
    A. a defective air governor
    B. a leak in a service brake chamber
    C. an air leak in the cab
    D. a defective air compressor
                                                                                    (D1.10)

6 Additional Test Questions for Practice

Additional Test Questions

Please note the letter and number in parentheses following each question. They match the task in Section 4 that discusses the relevant subject matter. You may want to refer to the overview using the cross-referencing key to help with questions posing problems for you.

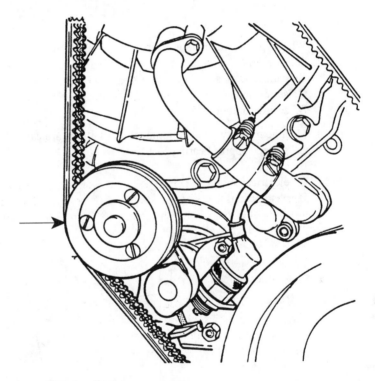

1. The item indicated by the arrow in the figure above is
   A. a water pump pulley.
   B. an air pump pulley.
   C. a fuel pump pulley.
   D. an idler pulley.

   (A1.3)

2. Which of the following is NOT an underside component of a trailer?
   A. bulkhead
   B. tires
   C. suspension
   D. landing gear

   (D4.10, D4.11, D5.1, D5.2)

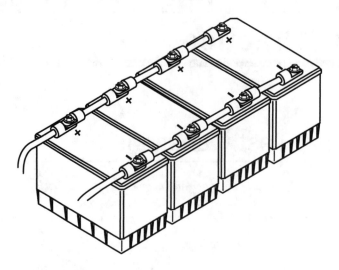

3. What type of connections are pictured in the figure?
   A. parallel
   B. parallel-series
   C. series
   D. series-parallel   (C1.2)

4. Most heavy-duty pneumatic fan clutch failures are caused by
   A. seized support bearings.
   B. air leaks.
   C. electrical shorts.
   D. obstructions.   (A4.1)

5. In the figure above, what could commonly be mounted above the roof depicted by arrow?
   A. identification lights
   B. solar panels
   C. skylight
   D. air vents for sleeper cab   (C3.2)

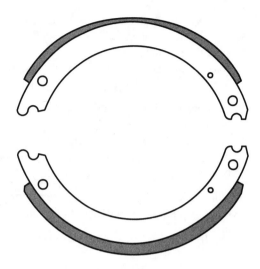

6. Which of these conditions do the shoes in the figure above show?
   A. unequal wear
   B. poor contact at center of linings
   C. lining material tapered
   D. normal wear   (D1.15)

7. Which of the following is NOT part of an automatic chassis lube system?
   A. zerk fittings
   B. reservoir
   C. pump
   D. distribution lines   (D1.19)

8. All of these could cause a drive axle failure of a single drive axle vehicle EXCEPT
   A. incorrect ratio.
   B. normal wear.
   C. lubrication failure.
   D. fatigue.   (D3.14, D3.15)

9. The major changes introduced by the Clean Air Act in heavy-duty truck air conditioning
   systems relate to
   A. certification of A/C technicians.
   B. maintaining older equipment.
   C. replacing leaking equipment.
   D. refrigerants and their handling.   (B4.1, B4.2)

10. A digital EST is connected to a truck databus during a routine PMI to read engine data.
    Technician A says active trouble codes cannot be erased. Technician B says that an event audit
    trail can be read. Who is correct?
    A. A only
    B. B only
    C. Both A and B
    D. Neither A nor B   (B1.7)

11. A digital diagnostic reader (DDR) is connected to the ATA connector during a routine PMI to
    read engine data. Technician A says active codes can be read. Technician B says that historic
    codes can be erased. Who is correct?
    A. A only
    B. B only
    C. Both A and B
    D. Neither A nor B   (B1.7)

12. The best way to clean a rear-axle breather is to
    A. replace the breather.
    B. soak the breather in gasoline.
    C. soak in solvent, then blow out with compressed air.
    D. use a rag to wipe the vent orifice clean.   (D3.12)

13. When checking for diesel engine fuel system leaks, the engine must
    A. be cold.
    B. be at normal operating temperature.
    C. be running.
    D. be off.   (A1.4)

14. Technician A says to connect the battery positive cable first when connecting a battery.
    Technician B says to disconnect the ground cable first when disconnecting a battery. Who
    is correct?
    A. A only
    B. B only
    C. Both A and B
    D. Neither A nor B   (C1.2)

15. Technician A says the Department of Transportation (DOT) does not require PMI records on
    lubrication. Technician B says that maintenance records should be referenced before
    undertaking major repairs. Who is correct?
    A. A only
    B. B only
    C. Both A and B
    D. Neither A nor B   (B3.8)

16. Which of these is the LEAST important check when performing a road test?
    A. verifying starter current draw
    B. verifying that air reservoir tanks are not leaking
    C. verifying that the tires are in good condition and properly inflated
    D. checking for any unusual noises during road test operation   (D5.1)

17. Technician A says that an inductive clamp ammeter could be used to check alternator output.
    Technician B says that alternator output decreases as temperature increases. Who is correct?
    A. A only
    B. B only
    C. Both A and B
    D. Neither A nor B   (C2.2)

18. Technician A says that all current tractor-trailers are required to have automatic slack adjusters.
    Technician B says that automatic slack adjusters increase the legal free stroke dimension. Who
    is correct?
    A. A only
    B. B only
    C. Both A and B
    D. Neither A nor B   (D1.17)

19. What is the most cited maintenance item on transport trailers?
    A. tires
    B. lights
    C. suspension
    D. brakes   (E8)

20. A tractor with air-spring suspension sits low when coupled to a trailer and sits high without the trailer. Technician A says a leaking air spring could be the problem. Technician B says an improperly adjusted height control valve could be the problem. Who is correct?
    A. A only
    B. B only
    C. Both A and B
    D. Neither A nor B   (D4.12)

21. If the floating caliper on a hydraulic brake truck is seized, which of these conditions will result?
    A. brake pedal fade
    B. brake pedal pulsations
    C. reduced braking force
    D. the brakes will grab   (D2.5)

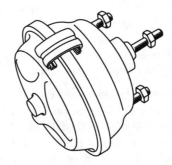

22. An audible air leak is observed that appears to be coming from the front service brake chamber in the figure above during an annual inspection. Which of the following is the LEAST-Likely cause?
    A. loose mounting bolts
    B. ruptured diaphragm
    C. loose chamber clamp
    D. missing clamp bolt   (D1.14)

23. When checking headlight aiming, Technician A says that the vehicle should be on level ground at a distance of 35 feet from a screen. Technician B says you must always use headlight aiming equipment when adjusting headlights. Who is correct?
    A. A only
    B. B only
    C. Both A and B
    D. Neither A nor B   (C3.2)

24. After a stretch of rough road is driven, the steering wheel continues to shake for a few seconds. Which of these could be the cause?
    A. leaking front shock absorbers
    B. low tire pressure
    C. rusted rear shock absorbers
    D. missing jounce bumper   (D4.11)

25. What would be the LEAST-Likely cause of a high amperage reading when checking starter current draw?
    A. short in the starter motor
    B. mechanical resistance due to binding engine
    C. starter wire gauge is too small
    D. short to ground in the starter wire   (C1.2, C1.5, C1.6)

26. In what position should the height control valve lever be in when a truck with an air spring suspension system is at its normal height and the valve is in its neutral position?
    A. about 45 degrees upward from level
    B. parallel to the ground
    C. about 45 degrees downward from level
    D. approximately vertical   (D4.12, D4.13)

27. All of the following would cause a constant leak from the air dryer purge valve EXCEPT
    A. foreign particles on valve seat.
    B. a purge valve seal damaged.
    C. a purge valve frozen.
    D. a restricted purge line.   (D1.3)

28. The driver complains about a rotten egg smell in the truck cab that is most noticeable when the A/C system is first turned on. The cause of this problem could be a
    A. leaking evaporator core.
    B. leaking condenser core.
    C. leaking heater core.
    D. plugged evaporator case drain.   (B4.3)

29. To eliminate a mildew smell from the A/C output air
    A. remove the evaporator case, clean it with a vinegar and water solution, and dry it thoroughly before reinstallation.
    B. spray a disinfectant into the panel outlets.
    C. pour a small amount of alcohol into the air intake plenum.
    D. place an automotive deodorizer under the dash.   (B4.4)

30. All of these conditions could be CVSA out-of-service criteria EXCEPT
    A. an exhaust leak under the sleeper.
    B. a wiring harness that contacts the exhaust.
    C. an inaccurate exhaust temperature gauge.
    D. a missing exhaust heat shield.   (A3.2)

31. Technician A says that multiple leaf springs should be replaced as an assembly. Technician B says that one end of semi-elliptic springs are free to move forward and backward. Who is correct?
    A. A only
    B. B only
    C. Both A and B
    D. Neither A nor B   (D4.10)

32. Technician A says that checking clutch operation and transmission shifting is part of a drive test during a PMI. Technician B claims that when a drive test is performed, road speed governor actual speed should be checked. Who is correct?
    A. A only
    B. B only
    C. Both A and B
    D. Neither A nor B   (E1, E5)

33. Which of the following is the standard fifth wheel plate height?
    A. 39 inches
    B. 43 inches
    C. 47 inches
    D. 49 inches   (D6.1, D6.4)

34. All of these spring types can be found on a heavy-duty truck suspension EXCEPT
    A. constant rate.
    B. progressive rate.
    C. auxiliary.
    D. coil.   (D4.10)

35. All of the following could cause the power-steering-pump pulley to become misaligned EXCEPT
    A. an over-pressed pulley.
    B. loose fit from pulley hub to pump shaft.
    C. a loose pump mounting bracket.
    D. a broken engine mount.   (A1.3, D4.2)

36. Technician A says that a circuit should be energized when using an ohmmeter. Technician B says that a voltage drop test is an accurate means of identifying high resistance in an energized circuit. Who is correct?
    A. A only
    B. B only
    C. Both A and B
    D. Neither A nor B   (C1.5, C2.2, C3.1)

37. Technician A says that a minor cooling system leak is sufficient criteria for a deadlining. Technician B says there must be sufficient vacuum in a vacuum assist brake system for there to be at least one full application after engine shut-off or it must be deadlined. Who is correct?
    A. A only
    B. B only
    C. Both A and B
    D. Neither A nor B   (A1.4, D2.6)

38. Technician A says that S-cam-actuated foundation brakes are the most common brake system on heavy-duty trucks today. Technician B says that slack adjusters and S-cams convert linear motion into brake torque. Who is correct?
    A. A only
    B. B only
    C. Both A and B
    D. Neither A nor B   (D1.18)

39. A worn front steering bushing can cause all of the following EXCEPT
    A. loose feeling in the steering wheel when driving straight.
    B. noise when hitting a bump in the road.
    C. excessive use of power steering fluid.
    D. pull to the left while driving.   (D4.1, D4.5, D4.12)

40. Technician A says the vibration damper counterbalances the back-and-forth twisting motion of the crankshaft each time a cylinder fires. Technician B says if the seal contact area on the vibration damper hub is scored, the damper assembly must be replaced. Who is correct?
    A. A only
    B. B only
    C. Both A and B
    D. Neither A nor B   (A1.2)

41. The diameter on a brake drum is greater at the edges of the friction surface than in the center. This drum condition is
    A. bell-mouth drum.
    B. concave brake drum.
    C. drum out-of-round.
    D. convex brake drum.   (D2.8)

42. What system could use a sliding gate valve?
    A. cooling system
    B. exhaust brake system
    C. air conditioning system
    D. electrical system   (E9)

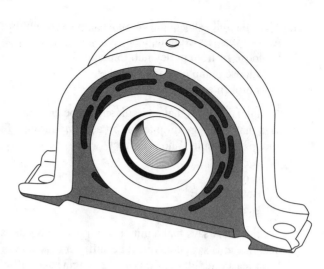

43. In the figure above, the component shown is
    A. a one-way clutch.
    B. a vertical yoke support.
    C. an exhaust pipe support.
    D. a center bearing support.   (D3.10)

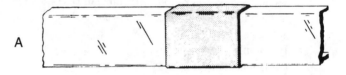

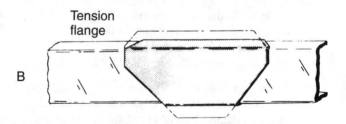

44. A technician is inspecting a frame rail as shown in the figure above. Technician A says the frame rail marked "A" is likely to result in frame failure. Technician B says the frame reinforcement on the frame rail marked "B"; is called an external channel reinforcement. Who is correct?
    A. A only
    B. B only
    C. Both A and B
    D. Neither A nor B   (D6.6)

45. When adjusting a clutch linkage, Technician A says that the pedal-free travel should be about 1½ to 2 inches (38.1 to 50.8 mm). Technician B says that release bearing to clutch brake travel should be less than ½ inch (12.7 mm). Who is correct?
    A. A only
    B. B only
    C. Both A and B
    D. Neither A nor B   (D3.4)

46. Technician A says that an engine can be checked for leaks while the engine is off. Technician B says to also check the engine for leaks while it is running. Who is correct?
    A. A only
    B. B only
    C. Both A and B
    D. Neither A nor B   (A1.4)

47. An inspection is being performed on a heavy-duty truck brake system and grease-soaked linings are found on three wheels of a trailer's rear tandem. Technician A says to inform the driver. Technician B writes up the problems and informs the driver. Who is correct?
    A. A only
    B. B only
    C. Both A and B
    D. Neither A nor B   (D1.15)

48. A technician observes a whitish, milky substance when changing the fluid in a drive axle. This condition is likely caused by
    A. normal condensation.
    B. infrequent driving and short trips.
    C. the axle being submerged in water.
    D. the vehicle being frequently driven during rainy or wet conditions.   (D3.15, D3.16)

49. With the brakes applied at 80 psi (551 kPa), leakage at the brake application valve exhaust port should not exceed
    A. a 2-inch (50.8 mm) bubble in 1 second.
    B. a 2-inch (50.8 mm) bubble in 2 seconds.
    C. a 3-inch (76.2 mm) bubble in 3 seconds.
    D. a 1-inch (25.4 mm) bubble in 3 seconds.   (D1.5)

50. While diagnosing an ABS with an ABS check switch and a light in the instrument panel
    A. when the ABS check switch and the ignition switch are turned on, the ABS light flashes blink codes.
    B. the first set of flashes in the blink code sequence indicates whether the system is ABS only or ABS and ATC.
    C. when there are no electrical defects in the ABS, a 000 blink code is revealed.
    D. a blink code indicates whether an electrical defect is in a wheel speed sensor or the connecting wiring.   (D1.12)

51. What type of fluid should be added to a hydraulic clutch master cylinder with a low fluid level during a PMI?
    A. power steering fluid
    B. DOT 3 brake fluid
    C. motor oil
    D. hydraulic fluid   (D3.3)

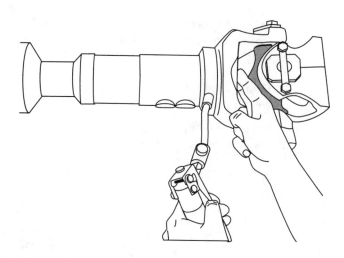

52. The procedure being performed in the figure above is
    A. lubricating the rear U-joint.
    B. lubricating the slip joint.
    C. lubricating the center bearing.
    D. lubricating the rear bearing.

(D3.13)

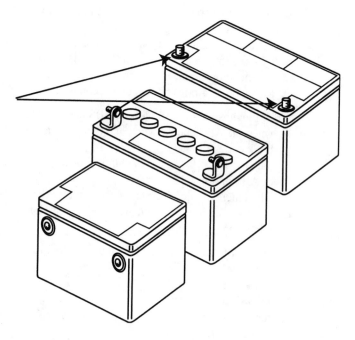

53. What type of battery do the arrows in the figure above indicate?
    A. top terminal, low maintenance
    B. stud terminal, maintenance free
    C. side terminal, maintenance free
    D. inverted terminal, high maintenance

(C1.2)

54. When winterizing a truck, which of the following procedures should be done first?
    A. replacing lost coolant
    B. cleaning the coolant
    C. adding antifreeze
    D. testing the condition of the coolant

(A4.7)

55. When inspecting a clutch linkage, a technician will check for
    A. mechanical advantage gained.
    B. linkage angles.
    C. clevis leverage.
    D. binding or wear.   (D3.2)

56. In addition to replacing the fuel filter and priming the fuel system, what else should be performed when doing a PMI on a fuel system?
    A. check the fuel tank mountings
    B. tests fuel pour point
    C. check heater lines for proximity to the fuel lines
    D. recalibrate the fuel gauge   (A2.1)

57. A turbocharger is being inspected for oil leaks. Technician A says to check it while the engine is running. Technician B measures turbine shaft radial runout with the engine running. Who is correct?
    A. A only
    B. B only
    C. Both A and B
    D. Neither A nor B   (A1.4, A3.4)

58. A driver complains of a door not closing properly and has brought it in for service. Technician A immediately greases the hinges. Technician B retrieves the vehicle records and checks for past problems with the doors and makes a note of the driver reports. Who is correct?
    A. A only
    B. B only
    C. Both A and B
    D. Neither A nor B   (B3.9)

59. Technician A says it is good practice to check underground fuel storage tanks for water content. Technician B says batteries can be kept from freezing if they have a full charge. Who is correct?
    A. A only
    B. B only
    C. Both A and B
    D. Neither A nor B   (A2.1, C1.3)

60. Technician A states that a vehicle should be deadlined if weld repairs are found to have been made on pintle chains and hooks. Technician B says that at least one functioning amber or red light on rear loads projecting more than 4 feet beyond the vehicle body is required. Who is correct?
    A. A only
    B. B only
    C. Both A and B
    D. Neither A nor B   (B2.2)

61. A heavy-duty diesel powered truck is having its electrical system inspected. The interior lights glow dimly even at a high engine speed. Technician A says that a defective ballast resistor should be suspected. Technician B says to check the cab ground integrity. Who is correct?
    A. A only
    B. B only
    C. Both A and B
    D. Neither A nor B   (C3.1)

62. In a heavy-duty truck, a pyrometer is used to indicate
    A. road speed.
    B. fuel pressure.
    C. exhaust gas temperature.
    D. the battery charging circuit.   (B1.2, B1.3)

63. What damage can overgreasing cause?
   A. channeling
   B. blowing out a bearing seal
   C. scoring
   D. accumulation of grit around greasing point   (D3.13)

64. Technician A says that the condition of the batteries should be verified with a load test prior to performing an alternator current output test. Technician B says that a load tester with an inductive pick-up may be used to test alternator current output. Who is correct?
   A. A only
   B. B only
   C. Both A and B
   D. Neither A not B   (C2.2, C2.3)

65. Technician A says that a short to ground in the wire from a gauge to the sending unit may cause the gauge to fail. Technician B says that most trucks equipped with electronic gauges perform a self-test when the ignition is first turned on. Who is correct?
   A. A only
   B. B only
   C. Both A and B
   D. Neither A nor B   (B1.3)

66. Technician A says that on the PMI you do not have to check the cab-raising hydraulic cylinders for leaks. Technician B says that on the PMI you must check to see if the cab will raise and lower properly. Who is correct?
   A. A only
   B. B only
   C. Both A and B
   D. Neither A nor B   (B3.10)

67. Visual inspection of the power steering belt(s) can reveal all the following EXCEPT
   A. proper belt tension.
   B. premature wear due to misalignment.
   C. correct orientation of dual belt application.
   D. proper belt seating in the pulley.   (A1.3, D4.2)

68. A combination vehicle consisting of a tractor, trailer, and a converter dolly is being inspected. Technician A says that the tractor must carry a copy of each inspection form in the cab. Technician B says that one inspection form is good for all trailers, dollies, and towed vehicles in the rig. Who is correct?
   A. A only
   B. B only
   C. Both A and B
   D. Neither A nor B   (B2.2)

69. A check of the general condition of the alternator is being performed. Technician A says to check for out-of-round mounting holes. Technician B tightens the belts before making performance tests. Who is correct?
   A. A only
   B. B only
   C. Both A and B
   D. Neither A nor B   (A1.3, C2.1)

70. Technician A says that an inspection of the road speed governor is required during the pre-trip inspection. Technician B says that more emphasis should be placed on the post-trip inspection rather than the pre-trip inspection. Who is correct?
    A. A only
    B. B only
    C. Both A and B
    D. Neither A nor B   (E5)

71. The cloud point of diesel fuel can be defined as
    A. the lowest temperature at which it will flow.
    B. the temperature at which a haze of wax crystals will form.
    C. the highest temperature before evaporation begins.
    D. the temperature at which the fuel produces black smoke.   (A2.1, A2.3)

72. A PMI is being performed on a heavy-duty truck. Technician A says one person can properly perform a steering system check. Technician B says not to adjust any belts before performing any tests. Who is correct?
    A. A only
    B. B only
    C. Both A and B
    D. Neither A nor B   (A1.3, D4.1)

73. A slack adjuster pushrod is being serviced on a vehicle following a driver tag on a post-trip inspection. Technician A says that the service performed must be logged before the vehicle can be put back into service. Technician B says the service performed should be recorded on a "B"; PMI maintenance sheet. Who is correct?
    A. A only
    B. B only
    C. Both A and B
    D. Neither A nor B   (D1.19)

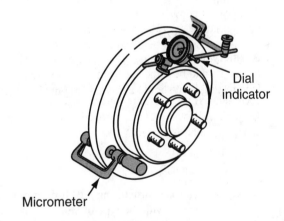

Dial indicator

Micrometer

74. What is being performed in the figure above?
    A. lateral runout and rotor thickness
    B. wheel bearing runout
    C. rotor thickness and radial runout
    D. parallelism   (D1.16)

75. The rear of the first trailer of a double trailer combination is being checked. Technician A says a cracked converter dolly chain is sufficient reason to deadline the unit. Technician B says that a wear reduction of more than 20 percent of the drawbar eye should deadline the unit. Who is correct?
    A. A only
    B. B only
    C. Both A and B
    D. Neither A nor B

    (B2.2, D6.4)

76. A cooling system on a heavy truck is being tested. Technician A says that the color of the coolant indicates whether it is EG, PG, or ELC. Technician B says to drain off a gallon of coolant before performing a pressure test. Who is correct?
    A. A only
    B. B only
    C. Both A and B
    D. Neither A nor B

    (A4.4, A4.7)

77. During a routine drive axle oil change, a technician observes a few metal particles on the magnetic plug of the drive axle. What should the technician do?
    A. Inform the customer that further investigation is needed.
    B. Inform the customer of the condition and instruct him to change oil in 30 days and reevaluate.
    C. Determine that a small amount of metal particles on the magnetic plug is normal.
    D. Disassemble the drive axle to locate the cause.

    (D3.15, D3.17)

78. A driveline with new driveshafts is being inspected for repetitive vibration related failures. Technician A says the problem is unlikely to be in the driveshaft since it is of identical size and construction to the original driveshaft. Technician B says the problem could be in the transmission because of high vibration encountered at high output shaft speeds. Who is correct?
    A. A only
    B. B only
    C. Both A and B
    D. Neither A nor B

    (D3.10)

79. A battery is being tested on a truck during PMI inspection. Technician A says that the surface charge should be removed from a battery before it can be load tested. Technician B says when load testing, you should apply a load equal to 1–2 of the batteries CCA rating for 15 seconds. Who is correct?
    A. A only
    B. B only
    C. Both A and B
    D. Neither A nor B

    (C1.3, C1.4)

80. A fleet manager is restructuring a PMI program. Technician A says that fire extinguisher distribution varies with the type of vehicle it is installed in. Technician B says that it is the fleet manager's responsibility to see that personnel performing annual inspections are qualified. Who is correct?
    A. A only
    B. B only
    C. Both A and B
    D. Neither A nor B

    (B2.2)

81. All of these conditions could cause high vibration in a driveshaft EXCEPT
    A. dents on the driveshaft tubing.
    B. binding U-joint of the crosses.
    C. burrs on the shaft flange surfaces that prevent it from seating.
    D. rust on the driveshaft tube.   (D3.10)

82. A gauge cluster is being inspected for proper function and illumination. Technician A says federal regulations forbid dispatching a vehicle that is likely to break down. Technician B says that a function check of the instrument panel is required under a PMI. Who is correct?
    A. A only
    B. B only
    C. Both A and B
    D. Neither A nor B   (B1.3)

7 Appendices

Answers to the Test Questions for the Sample Test Section 5

1. B	26. A	51. D	76. A
2. B	27. A	52. A	77. B
3. C	28. B	53. A	78. C
4. A	29. D	54. C	79. A
5. D	30. D	55. B	80. D
6. B	31. A	56. B	81. A
7. B	32. B	57. B	82. A
8. D	33. A	58. C	83. A
9. C	34. B	59. A	84. A
10. C	35. B	60. C	85. D
11. B	36. A	61. C	86. C
12. B	37. C	62. B	87. D
13. B	38. C	63. B	88. A
14. B	39. C	64. A	89. B
15. D	40. A	65. C	90. D
16. D	41. C	66. D	91. B
17. A	42. C	67. A	92. A
18. B	43. D	68. C	93. B
19. B	44. A	69. D	94. A
20. C	45. A	70. B	95. D
21. A	46. A	71. A	96. B
22. B	47. C	72. B	97. D
23. B	48. C	73. B	98. A
24. B	49. C	74. C	99. C
25. B	50. C	75. B	100. D

101.	A	**117.**	D	**132.**	A	**147.**	D
102.	A	**118.**	A	**133.**	D	**148.**	A
103.	A	**119.**	C	**134.**	C	**149.**	C
104.	D	**120.**	C	**135.**	A	**150.**	C
105.	C	**121.**	D	**136.**	B	**151.**	B
106.	C	**122.**	B	**137.**	D	**152.**	B
107.	B	**123.**	C	**138.**	A	**153.**	B
108.	C	**124.**	C	**139.**	A	**154.**	B
109.	B	**125.**	D	**140.**	A	**155.**	D
110.	C	**126.**	B	**141.**	C	**156.**	A
111.	A	**127.**	C	**142.**	C	**157.**	C
112.	D	**128.**	C	**143.**	D	**158.**	C
113.	B	**129.**	A	**144.**	A	**159.**	D
114.	C	**130.**	D	**145.**	A	**160.**	B
115.	C	**131.**	D	**146.**	C	**161.**	D
116.	A						

Explanations to the Answers for the Sample Test Section 5

Question #1
Answer A is wrong. Overinflation causes cracks in the rim base.
Answer B is correct. Overloading causes lug hole cracking.
Answer C is wrong. This causes cracks in the rim base.
Answer D is wrong. This generally does not cause cracks.

Question #2
Answer A is wrong. A cooling system tester is hooked up to the inlet spout of the radiator.
Answer B is correct. A belt tension gauge is shown in the figure.
Answer C is wrong. A vacuum gauge uses a small vacuum hose for hook-up.
Answer D is wrong. A dial indicator does not have a handle but does have a small pointer.

Question #3
Answer A is wrong. A water temperature gauge would have different values and symbol(s).
Answer B is wrong. An oil temperature gauge would have different values and symbol(s).
Answer C is correct. An oil pressure gauge is shown.
Answer D is wrong. A transmission temperature gauge would have different values and symbols.

Question #4
Answer A is correct. A solenoid valve activates the pump.
Answer B is wrong. There are no relays in an air-driven automatic chassis lubrication system.
Answer C is wrong. There is no ECM to control an ACLS.
Answer D is wrong. A toggle switch would make the system manual.

Question #5
Answer A is wrong. Wheel hub runout is not being checked.
Answer B is wrong. There is no aligning a wheel hub to an axle hub.
Answer C is wrong. Wheel bearings are usually installed by hand.
Answer D is correct. The figure shows the installation of front wheel bearing seals.

Question #6
Answer A is wrong. Up to ⅛ inch is acceptable.
Answer B is correct. ⅛ inch play is acceptable free-play.
Answer C is wrong. ¼ inch is too much play.
Answer D is wrong. ½ inch is too much play.

Question #7
Answer A is wrong. Vertical play is checked with the indicator on top of the knuckle.
Answer B is correct. A knuckle upper bushing play measurement is shown in the figure.
Answer C is wrong. The indicator should be mounted on the lower bushing area.
Answer D is wrong. There is no kingpin axial play test.

Question #8
Answer A is wrong. A hydraulic liftgate cylinder would be much larger.
Answer B is wrong. The indicated object is not a solenoid.
Answer C is wrong. The kingpin does not have an automatic locking relay.
Answer D is correct. A hydraulic spring parking brake actuator is shown in the figure.

Question #9
Answer A is wrong. The water does not remain mixed with the fuel.
Answer B is wrong. Water does not float on top of the fuel.
Answer C is correct. The water settles to the bottom of the fuel tank.
Answer D is wrong. The water usually does not flow through the filters.

Question #10
Answer A is wrong. Technician B is also correct.
Answer B is wrong. Technician A is also correct.
Answer C is correct. Both technicians are correct. A loose grab handle may cause personal injury. The steps must be level and free of debris.
Answer D is wrong. Both technicians are correct.

Question #11
Answer A is wrong. 1,000 rpm would usually be too slow for maximum alternator output.
Answer B is correct. Alternator output is usually tested at 1,500–2,000 rpm.
Answer C is wrong. The alternator will not produce maximum output at idle speed.
Answer D is wrong. An alternator produces maximum output at 1,500–2,000 rpm.

Question #12
Answer A is wrong. Loose tie-rod ends do not cause a binding steering wheel.
Answer B is correct. Loose or worn steering shaft U-joints may cause a binding steering wheel.
Answer C is wrong. Loose drag-link ends do not cause a binding steering wheel.
Answer D is wrong. Worn kingpin bushings do not cause a binding steering wheel.

Question #13
Answer A is wrong. The oil level will increase as the temperature increases.
Answer B is correct. An increase in oil temperature will raise the oil level in the transmission.
Answer C is wrong. The oil level will increase as the temperature increases.
Answer D is wrong. The oil level will increase as the temperature increases.

Question #14
Answer A is wrong. A hydraulic wheel cylinder may occasionally need to be rebored.
Answer B is correct. A wheel cylinder should always be inspected on a PM.
Answer C is wrong. A wheel cylinder does not need replacing periodically, but needs servicing on occasion.
Answer D is wrong. Wheel cylinders are generally constructed of cast iron.

Question #15
Answer A is wrong. Safety triangles are the most likely item found in a safety kit.
Answer B is wrong. Safety flares are the second most likely item found in a safety kit.
Answer C is wrong. A fire extinguisher is generally included in a safety kit.
Answer D is correct. A spare tool box is not found among the safety items required in the cab by the CVSA standard.

Question #16
Answer A is wrong. Grease lines serve as avenues for the grease to travel from one place to another.
Answer B is wrong. Reservoirs are the components that store the medium until needed.
Answer C is wrong. The pump is the delivery mechanism for the system.
Answer D is correct. The manifold or distribution block is not part of the automatic chassis lube system.

Question #17
Answer A is correct. Excessive brake pedal free play causes a low, firm pedal.
Answer B is wrong. Air in the hydraulic system causes a low, spongy brake pedal.
Answer C is wrong. Brake fluid contaminated with moisture causes a low, spongy brake pedal.
Answer D is wrong. Low brake fluid level causes a low, spongy brake pedal.

Question #18
Answer A is wrong. Excessive kingpin causes front wheel shimmy.
Answer B is correct. Steering gear relief plungers are the least likely cause of a front wheel shimmy because they are only a factor at steering lock.
Answer C is wrong. Looseness in steering linkages can cause wheel shimmy.
Answer D is wrong. Wheel shimmy can be caused by tire imbalance.

Question #19
Answer A is wrong. Alternator amperage output is not being indicated because an ammeter is not being used.
Answer B is correct. The battery is being load tested.
Answer C is wrong. Voltmeter #1 is not indicating field circuit voltage drop.
Answer D is wrong. Voltmeter #2 is not indicating voltage drop of the battery negative circuit.

Question #20
Answer A is wrong. The bearings in the figure are lubricated with the specified oil.
Answer B is wrong. The bearings are not permanently lubricated with the specified oil.
Answer C is correct. The bearings require a specific hub end play.
Answer D is wrong. The bearings only have an inner hub seal.

Question #21
Answer A is correct. Only Technician A is correct. The back-up warning system provides a warning to anyone standing behind the truck.
Answer B is wrong. The back-up warning system provides an audible warning.
Answer C is wrong. Only Technician A is correct.
Answer D is wrong. Only Technician A is correct.

Question #22
Answer A is wrong. Air seat spring suspensions do not eliminate the need for cab shock absorbers.
Answer B is correct. Only Technician B is correct. You can adjust the leveling valve to maintain the proper cab height.
Answer C is wrong. Only Technician B is correct.
Answer D is wrong. Only Technician B is correct.

Question #23
Answer A is wrong. Air seat spring suspensions DO NOT eliminate the need for cab shock absorbers.
Answer B is correct. Only Technician B is correct. You can adjust the leveling valve to maintain the proper seat height.
Answer C is wrong. Only Technician B is correct.
Answer D is wrong. Only Technician B is correct.

Question #24
Answer A is wrong. The transmission oil sample must be taken mid-way through the drain-off.
Answer B is correct. Only Technician B is correct. You always take a transmission oil sample mid-way through the drain-off.
Answer C is wrong. Only Technician B is correct.
Answer D is wrong. Only Technician B is correct.

Question #25
Answer A is wrong. 200,000 miles is not a suitable interval for changing axle lubricant.
Answer B is correct. 100,000 miles is the most likely change interval for class "A" highway vehicles.
Answer C is wrong. 50,000 miles is not a suitable interval for changing axle lubricant.
Answer D is wrong. 25,000 miles is not enough mileage for an axle lubricant change.

Question #26

Answer A is correct. The technician is performing a hydraulic drum brake adjustment.
Answer B is wrong. The technician is performing a hydraulic drum brake adjustment.
Answer C is wrong. The technician is performing a hydraulic drum brake adjustment.
Answer D is wrong. The technician is performing a hydraulic drum brake adjustment.

Question #27

Answer A is correct. Only Technician A is correct. Moisture and corrosion on battery tops causes faster battery self-discharge.
Answer B is wrong. Excessive corrosion on the battery box and cover indicates overcharging.
Answer C is wrong. Only Technician A is correct.
Answer D is wrong. Only Technician A is correct.

Question #28

Answer A is wrong. Aerated fluid will appear with small bubbles trapped within it.
Answer B is correct. Dirt or rust contaminated brake fluid will appear cloudy or opaque.
Answer C is wrong. Burned or otherwise overheated brake fluid will have a distinct and pungent odor of being unlike any other brake fluid, such that a distinction can be made.
Answer D is wrong. Brake fluid will not freeze above the working limits of a vehicle.

Question #29

Answer A is wrong. Some evaporator filters are made of paper or a metallic mesh.
Answer B is wrong. Evaporator filters are not designed to remove moisture from the air.
Answer C is wrong. Evaporator filters are replaceable.
Answer D is correct. They are designed to remove dust and dirt from cab and sleeper air.

Question #30

Answer A is wrong. A tractor protection valve check is always part of an air brake system check.
Answer B is wrong. Proper operation of the spring brake inversion system is vital.
Answer C is wrong. Nonregenerative desiccant packs must be replaced regularly.
Answer D is correct. Checking the operation of the clutch brake is not part of the brakes. It is in the clutch unit.

Question #31

Answer A is correct. A restricted air intake may cause black smoke in the exhaust.
Answer B is wrong. Misfiring injectors that do not inject fuel do not cause black smoke in the exhaust.
Answer C is wrong. Coolant entering the combustion chambers causes white or gray smoke in the exhaust.
Answer D is wrong. Oil entering the combustion chambers causes blue smoke in the exhaust.

Question #32

Answer A is wrong. Most PM records on heavy trucks are required by the DOT.
Answer B is correct. Only Technician B is correct. Maintenance records are an invaluable source of information in the life of a heavy truck.
Answer C is wrong. Only Technician B is correct.
Answer D is wrong. Only Technician B is correct.

Question #33

Answer A is correct. Only Technician A is correct. The operation of the road speed governor is checked during the road/operational test.
Answer B is wrong. A post-trip inspection is of secondary importance.
Answer C is wrong. Only Technician A is correct.
Answer D is wrong. Only Technician A is correct.

Question #34
Answer A is wrong. Clutch pedal free travel should be about 1.5 to 2 inches.
Answer B is correct. Only Technician B is correct. The last inch of clutch-pedal travel activates the clutch brake on trucks equipped with a clutch brake.
Answer C is wrong. Only Technician B is correct.
Answer D is wrong. Only Technician B is correct.

Question #35
Answer A is wrong. A concave drum would be bowed out, not in. The concave condition indicates drum diameter greater at the friction surface center as compared to the edges of the friction area.
Answer B is correct. A convex drum condition is shown. The convex condition indicates a drum diameter greater at the friction surface edges as compared to the center of the friction surface.
Answer C is wrong. An offset is not a condition.
Answer D is wrong. A shifted drum is not a condition.

Question #36
Answer A is correct. Loose mounting bolts would cause the chamber to move when the brakes are applied, but would be unlikely to cause an audible leak.
Answer B is wrong. A ruptured diaphragm could cause an air leak.
Answer C is wrong. A loose chamber clamp could cause an air leak.
Answer D is wrong. A missing clamp bolt could cause an air leak.

Question #37
Answer A is wrong. Technician B is also correct.
Answer B is wrong. Technician A is also correct.
Answer C is correct. Both technicians are correct. The alternator should be checked for out-of-round mounting holes. Alternator belts should be checked for proper tightness before making performance tests.
Answer D is wrong. Both technicians are correct.

Question #38
Answer A is wrong. Technician B is also correct.
Answer B is wrong. Technician A is also correct.
Answer C is correct. Both technicians are correct. You use Teflon lubricant on the door latches and dry graphite on the door locks. You mount the alcohol evaporator downstream from the air dryer.
Answer D is wrong. Both technicians are correct.

Question #39
Answer A is wrong. Technician B is also correct.
Answer B is wrong. Technician A is also correct.
Answer C is correct. Both technicians are correct. You turn the key on the ignition to perform electronic system self-checks. The ignition switch powers up most computer controlled chassis systems.
Answer D is wrong. Both technicians are correct.

Question #40
Answer A is correct. Only Technician A is correct. The air passages through the air-to-air intercooler should be checked for restrictions.
Answer B is wrong. The air-to-air intercooler does not use refrigerant to lower intake air temperature.
Answer C is wrong. Only Technician A is correct.
Answer D is wrong. Only Technician A is correct.

Question #41
Answer A is wrong. Technician B is also correct.
Answer B is wrong. Technician A is also correct.
Answer C is correct. Both technicians are correct. ⅜ inch is the maximum allowable when checking fan clutch play. Door latches and door locks should be lubricated when an inspection is performed.
Answer D is wrong. Both technicians are correct.

Question #42
Answer A is wrong. Technician B is also correct.
Answer B is wrong. Technician A is also correct.
Answer C is correct. Both technicians are correct. Overfilling a manual transmission could result in overheating of the transmission. Overfilling a manual transmission could cause excessive aeration of the transmission fluid.
Answer D is wrong. Both technicians are correct.

Question #43
Answer A is wrong. Using ether on glow plug equipped engines can cause serious internal engine damage.
Answer B is wrong. Ether is generally used only in cold weather conditions and should not be needed in warmer weather.
Answer C is wrong. Both technicians are wrong.
Answer D is correct. Neither technician is correct. Ether should never be used on a glow plug equipped engine. Ether is generally used only in cold weather conditions.

Question #44
Answer A is correct. Only Technician A is correct. You must inflate mounted tires in a safety cage or using a portable lock ring guard.
Answer B is wrong. You do not mount the tire on the truck before inflation because an improperly fitted rim ring could dislodge, causing injury.
Answer C is wrong. Only Technician A is correct.
Answer D is wrong. Only Technician A is correct.

Question #45
Answer A is correct. Only Technician A is correct. When performing a brake overhaul on a hydraulic rear drum brake system, you replace the return springs.
Answer B is wrong. The maximum allowable out-of-round specification on a brake drum is 0.015 inch.
Answer C is wrong. Only Technician A is correct.
Answer D is wrong. Only Technician A is correct.

Question #46
Answer A is correct. Only Technician A is correct. The device in the figure is to restrain the tire while it is being inflated.
Answer B is wrong. Once a tire's shape has been altered (belt has shifted, high speed distortion, etc.), there is no way to reshape the rubber structure back to its original form.
Answer C is wrong. Only Technician A is correct.
Answer D is wrong. Only Technician A is correct.

Question #47
Answer A is wrong. Technician B is also correct.
Answer B is wrong. Technician A is also correct.
Answer C is correct. Both technicians are correct. A defective water pump bearing may cause a growling noise with the engine idling. The water pump bearing may be ruined by coolant leaking past the water pump seal.
Answer D is wrong. Both technicians are correct.

Question #48
Answer A is wrong. Technician B is also correct.
Answer B is wrong. Technician A is also correct.
Answer C is correct. Both technicians are correct. A circle inspection includes items such as torn mud flaps, leaking hoses, and outdated permits. Performing a circle inspection first before any other inspection is part of a PMI.
Answer D is wrong. Both technicians are correct.

Question #49
Answer A is wrong. The height control valve arm may be moved upward to raise the suspension height.
Answer B is wrong. Wood locating pins are installed in the height control valve arm.
Answer C is correct. The air brake system pressure must be 70 psi (483 kPa) or more.
Answer D is wrong. The truck must be unloaded or empty.

Question #50
Answer A is wrong. Excessive toe-in will not cause a cut in the sidewall. Improper toe adjustment causes a feathered tire tread wear pattern.
Answer B is wrong. Camber misadjustment will not cause a cut in the sidewall. Improper camber causes wear on one side of the tire tread.
Answer C is correct. Careless driving is the most likely cause of a cut in the sidewall. Sidewall damage can lead to future tire failure.
Answer D is wrong. Caster will not cause a cut in the sidewall.

Question #51
Answer A is wrong. The indication is coolant in the oil, and a cracked block or cylinder head causes this condition.
Answer B is wrong. The indication is coolant in the oil, and a blown head gasket causes this condition.
Answer C is wrong. The indication is water in the oil, and a leaky oil cooler causes this condition.
Answer D is correct. A worn oil pump is the least likely cause of water droplets or gray oil on the dipstick because there is no connection between the cooling system and the oil pump.

Question #52
Answer A is correct. Refrigerant leaks at the condenser line fittings are indicated by an oily film on the fittings.
Answer B is wrong. Refrigerant leaks at the condenser line fittings are indicated by an oily film on the fittings.
Answer C is wrong. Refrigerant leaks at the condenser line fittings are indicated by an oily film on the fittings.
Answer D is wrong. Refrigerant leaks at the condenser line fittings are indicated by an oily film on the fittings.

Question #53
Answer A is correct. An EGR valve stuck open causes rough idle operation.
Answer B is wrong. A burned exhaust valve causes low compression.
Answer C is wrong. A valve adjustment that is too tight causes low cylinder compression.
Answer D is wrong. A bent intake valve causes low cylinder compression.

Question #54
Answer A is wrong. Having an open circuit in the sending circuit would cause a low temperature gauge reading.
Answer B is wrong. A short to the power feed circuit would open (blow) the fuse protecting it.
Answer C is correct. A short circuit in the sending unit could cause a continuous high temperature reading.
Answer D is wrong. An open or burned out gauge will cause no operation of that gauge.

Question #55
Answer A is wrong. A shock absorber outer tube touching the fluid reservoir will not cause premature bushing wear.
Answer B is correct. The shock absorber outer tube touching the fluid reservoir will cause a scraping noise during shock travel.
Answer C is wrong. A shock absorber outer tube touching the fluid reservoir will not cause main shaft seal failure.
Answer D is wrong. A shock absorber outer tube touching the fluid reservoir will not cause excessive fatigue in the mount assembly.

Question #56
Answer A is wrong. Never compress the joint with a C-clamp or large pliers.
Answer B is correct. Only Technician B is correct. Simply pushing against the joint with force that you can create with your hands should be enough to identify excessive wear.
Answer C is wrong. Only Technician B is correct.
Answer D is wrong. Only Technician B is correct.

Question #57
Answer A is wrong. When filling the cooling system, the air bleeder valves should be open.
Answer B is correct. Only Technician B is correct. If the coolant solution contains over 67 percent antifreeze, heat transfer from the cylinders to the coolant is adversely affected.
Answer C is wrong. Only Technician B is correct.
Answer D is wrong. Only Technician B is correct.

Question #58
Answer A is wrong. Technician B is also correct.
Answer B is wrong. Technician A is also correct.
Answer C is correct. Both technicians are correct. A cracked fan blade may cause a clicking or creaking noise with the engine idling. You should not stand or position your body in line with the fan while increasing the engine speed.
Answer D is wrong. Both technicians are correct.

Question #59
Answer A is correct. Only Technician A is correct. Activating the parking brake to brake the truck very quickly at low speeds will determine if the seat belt inertia reel locks.
Answer B is wrong. All seat belts and restraints of any type must be checked on a PMI.
Answer C is wrong. Only Technician A is correct.
Answer D is wrong. Only Technician A is correct.

Question #60
Answer A is wrong. Technician B is also correct.
Answer B is wrong. Technician A is also correct.
Answer C is correct. Both technicians are correct. The outlet temperature should be within several degrees of the set temperature on a truck equipped with automatic temperature control. The blower motor should operate in all speed positions.
Answer D is wrong. Both technicians are correct.

Question #61
Answer A is wrong. Technician B is also correct.
Answer B is wrong. Technician A is also correct.
Answer C is correct. Both technicians are correct. A clogged filter could cause the steering to become harder gradually. A loose steering gear mount could be binding the steering column.
Answer D is wrong. Both technicians are correct.

Question #62
Answer A is wrong. The charging circuit is not related to the starting circuit.
Answer B is correct. The cranking circuit is involved in a starter draw test. The starter draw test is used to measure the amount of current in amperes that the starter circuit draws to crank the engine. This information is used to determine the next step needed to pinpoint a starter system problem.
Answer C is wrong. The regulator circuit is not related to the starter circuit.
Answer D is wrong. The control circuit is not related to the starter circuit.

Question #63
Answer A is wrong. All tractors must have a functioning tractor protection valve.
Answer B is correct. Radial tires must have less than $\frac{1}{32}$ inch tread when measured in any 2 adjacent major tread grooves before it can be deadlined.
Answer C is wrong. Any movement of a steering pivot point under load is deadlined immediately.
Answer D is wrong. All air system support brackets must be in place.

Question #64

Answer A is correct. Only Technician A is correct. The hydraulic assist back-up system and warning devices should be checked at each inspection.
Answer B is wrong. The hydraulic assist back-up system should be checked at each inspection.
Answer C is wrong. Only Technician A is correct.
Answer D is wrong. Only Technician A is correct.

Question #65

Answer A is wrong. A visual inspection for cracks is a must.
Answer B is wrong. A visual inspection of the gasket surface will help ensure a tight leak-free seal.
Answer C is correct. Magna fluxing of housing assembly and metal parts to detect small cracks would typically not be performed.
Answer D is wrong. A visual inspection of all passageways is necessary to ensure oil flows.

Question #66

Answer A is wrong. Servicing the window personally does not fall under the federal regulations.
Answer B is wrong. Servicing the window personally does not fall under the federal regulations.
Answer C is wrong. Informing the service department will not accomplish any service on the vehicle.
Answer D is correct. You do record the observed problem on a PMI.

Question #67

Answer A is correct. When brake drums and shoes are excessively worn, trying to adjust them to specs will put the cam in a position to roll over when the brakes are applied.
Answer B is wrong. Application of air pressure affects all vehicle brakes and will not cause the S-cam to rollover.
Answer C is wrong. Improperly installed slack adjusters will not cause S-cam rollover.
Answer D is wrong. Brake chamber push rod length will not cause S-cam rollover.

Question #68

Answer A is wrong. Technician B is also correct.
Answer B is wrong. Technician A is also correct.
Answer C is correct. Both technicians are correct. A circle inspection includes items such as torn mud flaps, leaking hoses, and outdated permits to name a few. A technician performs a circle inspection first before any other inspection as part of a PMI.
Answer D is wrong. Both technicians are correct.

Question #69

Answer A is wrong. Turbocharger bearings that are starting to fail may caused reduced boost pressure.
Answer B is wrong. Improper wastegate adjustment may cause reduced boost pressure.
Answer C is wrong. A damaged compressor wheel may cause reduced boost pressure.
Answer D is correct. Turbocharger oil seals leaking oil into the exhaust cause blue smoke in the exhaust, but this problem does not cause reduced boost pressure.

Question #70

Answer A is wrong. Three floor cross members have to be broken and sagging below the lower rail for the vehicle to be tagged out of service.
Answer B is correct. Only Technician B is correct. If a drop frame is visibly twisted, the vehicle should be placed out of service.
Answer C is wrong. Only Technician B is correct.
Answer D is wrong. Only Technician B is correct.

Question #71

Answer A is correct. When the breather is plugged, internal pressure can build inside the transmission and push the oil out.

Answer B is wrong. The shift cover is an exit route for the pressurized oil, not the source of the pressure.

Answer C is wrong. The release bearing is not an exit route for pressurized oil, nor is it the source of the pressure.

Answer D is wrong. The rear seal is an exit route for the pressurized oil, not the source of the pressure.

Question #72

Answer A is wrong. It is not necessary to check an exhaust system while it is hot.

Answer B is correct. Only Technician B is correct. A prybar can used to check the looseness of the piping and the strength of the brackets when performing an exhaust system PMI.

Answer C is wrong. Only Technician B is correct.

Answer D is wrong. Only Technician B is correct.

Question #73

Answer A is wrong. Generally, if any components or systems are suspected of requiring servicing, those nonchecklisted areas are more thoroughly inspected and/or serviced immediately.

Answer B is correct. Only Technician B is correct. You always check the cargo ramps on a vehicle equipped with them during a PMI.

Answer C is wrong. Only Technician B is correct.

Answer D is wrong. Only Technician B is correct.

Question #74

Answer A is wrong. Technician B is also correct.

Answer B is wrong. Technician A is also correct.

Answer C is correct. Both technicians are correct. A malfunctioning oil pressure gauge is not a valid reason for tagging a vehicle out of service. A proper function check of the instrumentation panel is required under a PMI.

Answer D is wrong. Both technicians are correct.

Question #75

Answer A is wrong. If a 45-degree elbow in an air brake line is replaced with a 90-degree elbow, brake timing is adversely affected.

Answer B is correct. Only Technician B is correct. A smaller than specified air line slows the brake application.

Answer C is wrong. Only Technician B is correct.

Answer D is wrong. Only Technician B is correct.

Question #76

Answer A is correct. Only Technician A is correct. You must remove the wheel in most cases to measure pad thickness.

Answer B is wrong. If the brakes were applied shortly before measuring the linging-to-rotor clearance, this clearance will be LESS than specified.

Answer C is wrong. Only Technician A is correct.

Answer D is wrong. Only Technician A is correct.

Question #77
Answer A is wrong. Restricted radiator tubes may cause excessive coolant level in the recovery reservoir.
Answer B is correct. A thermostat that is stuck open does not cause excessive coolant level in the recovery reservoir.
Answer C is wrong. A blown head gasket could allow combustion chamber gases into the coolant, which would raise the coolant level in the recovery reservoir.
Answer D is wrong. An inoperative air-operated cooling fan clutch may cause excessive coolant level in the recovery reservoir.

Question #78
Answer A is wrong. A technician should never ignore a problem when a potential repair(s) poses a problem to the serviceability of the truck.
Answer B is wrong. Logging in an A/C serviceable problem is not the minimum responsibility of a technician.
Answer C is correct. The technician logs the problem, then informs the owner/driver that this is in violation of the Clean Air Act.
Answer D is wrong. A vehicle should never be repaired without proper paperwork in order.

Question #79
Answer A is correct. You do record the observed problem on a PMI.
Answer B is wrong. Informing the service department will not accomplish any service on the vehicle.
Answer C is wrong. Servicing the bumper personally does not fall under the federal regulations.
Answer D is wrong. Servicing the bumper personally does not fall under the federal regulations.

Question #80
Answer A is wrong. The pressure drop on a straight truck is 4 psi in two minutes.
Answer B is wrong. The brake pedal must be held down for the two minute test.
Answer C is wrong. Neither technician is correct.
Answer D is correct. Neither technician is correct. The pressure drop on a straight truck is 4 psi in two minutes. The brake pedal must be held down for the two minute test.

Question #81
Answer A is correct. Only Technician A is correct. When 25 in. Hg. air restriction is reached the element should be charged.
Answer B is wrong. Federal Motor Carrier Safety Regulations are very clear on confirming that a vehicle that is used commercially or industrially must be in a safe operating condition at all times, regardless of the cost to keep a vehicle in such a condition.
Answer C is wrong. Only Technician A is correct.
Answer D is wrong. Only Technician A is correct.

Question #82
Answer A is correct. The tractor protection valve closes to protect the tractor air supply if the supply line pressure drops below 20 psi (138 kPa).
Answer B is wrong. The tractor protection valve closes to protect the tractor air supply if the supply line pressure drops below 20 psi (138 kPa).
Answer C is wrong. The tractor protection valve closes to protect the tractor air supply if the supply line pressure drops below 20 psi (138 kPa).
Answer D is wrong. The tractor protection valve closes to protect the tractor air supply if the supply line pressure drops below 20 psi (138 kPa).

Question #83

Answer A is correct. Only Technician A is correct. A cooling system pressure reading of 15 psi is normal for pressure testing.

Answer B is wrong. A cooling system pressure tester does not generally require a normal engine operating temperature for operation. A cooling system pressure tester has a manually operated pump to raise system pressure without raising its temperature.

Answer C is wrong. Only Technician A is correct.

Answer D is wrong. Only Technician A is correct.

Question #84

Answer A is correct. You do move the handle to the fully applied position and record the air pressure.

Answer B is wrong. The air system does not have to be drained to test the trailer control valve.

Answer C is wrong. As long as the trailer brakes are fully charged, the initial gauge reading is unnecessary.

Answer D is wrong. Air leakage around the control handle should not occur.

Question #85

Answer A is wrong. A burned fusible link must be replaced with a length of specified fusible link wire.

Answer B is wrong. Circuit breakers can be reset.

Answer C is wrong. Neither technician is correct.

Answer D is correct. Neither technician is correct. A burned fusible link must be replaced with a length of the specified fusible link wire. A circuit breaker resets itself or can be reset and does not have to be replaced after repairing a short in a circuit.

Question #86

Answer A is wrong. Technician B is also correct.

Answer B is wrong. Technician A is also correct.

Answer C is correct. Both technicians are correct. Trucks equipped with power steering filters should have them changed, if required, as part of preventive maintenance inspections (PMI). When there has been a component failure, the power steering filter may be filled with debris from the failure, and should be changed.

Answer D is wrong. Both technicians are correct.

Question #87

Answer A is wrong. 100°F is too low.

Answer B is wrong. 140°F is too low.

Answer C is wrong. 90°F is too low.

Answer D is correct. 175°F is the correct value. Most OEM truck manufacturers recommend checking the power steering fluid level at an operating or working temperature of 175°F (79°C). With the engine at 1,000 rpm or less, turn the steering wheel slowly and completely in each direction several times to raise the fluid temperature. Check the reservoir for foaming as a sign of aerated fluid. The fluid level in the reservoir should be at the hot full mark on the dipstick.

Question #88

Answer A is correct. The vehicle is not tagged out of service unless wear in the pintle hook horn exceeds 20 percent of the horn section.

Answer B is wrong. The vehicle is tagged out of service if there are loose or missing pintle hook mounting bolts.

Answer C is wrong. The vehicle is tagged out of service if there are any welded repairs to the drawbar eye.

Answer D is wrong. The vehicle is tagged out of service if there are cracks in the pintle hook-mounting surface.

Question #89
Answer A is wrong. A defective cruise control module is not the most likely cause of improper cruise set speed.
Answer B is correct. A defective vehicle speed sensor (VSS) is the most likely cause of improper cruise control set speed.
Answer C is wrong. Improper size of tires on the drive axles is not the most likely cause of improper cruise control set speed.
Answer D is wrong. A defective cruise control switch is not the most likely cause of improper cruise control set speed.

Question #90
Answer A is wrong. Each vehicle must carry a copy of the inspection form attached to the vehicle with a copy of the report on file.
Answer B is wrong. Each vehicle must carry a copy of the inspection form attached to the vehicle with a copy of the report on file.
Answer C is wrong. Neither technician is correct.
Answer D is correct. Neither technician is correct. Each vehicle must carry a copy of the inspection form attached to the vehicle with a copy of the report on file.

Question #91
Answer A is wrong. The ABS warning light is illuminated to indicate an electrical defect in the system when the vehicle speed is above 4 mph (6.43 km/h).
Answer B is correct. Only Technician B is correct. The ABS warning light is illuminated to indicate an electrical defect in the system when the vehicle speed is above 4 mph (6.43 km/h).
Answer C is wrong. Only Technician B is correct.
Answer D is wrong. Only Technician B is correct.

Question #92
Answer A is correct. The clearance light is shown and mounted on the extreme right or left side of the cab. Federal lighting regulations require all vehicles 80 inches or wider to have two clearance lamps at the front and two at the rear at the widest point and as high as practicable.
Answer B is wrong. Identification lights are in the center of the roof of the cab. Federal lighting regulations for the front of vehicles 80 inches or wider require three identification lights spaced 6 inches–12 inches on center as high as practicable.
Answer C is wrong. Fog lights are usually mounted as close to the ground as possible, usually on the bumper.
Answer D is wrong. Intermediate lights do not exist.

Question #93
Answer A is wrong. 80W-90 gear lube is typically used in drive axles.
Answer B is correct. The top surface of the fifth wheel must be lubricated with a water-resistant lithium-based grease. The consistency of grease, a measure of its relative hardness, is commonly expressed in terms of ASTM penetration or NLGI (National Lubricating Grease Institute) consistency number. Lubrication of the fifth-wheel top plate may be done by direct application or through the grease fittings on the underside of the top plate.
Answer C is wrong. Bearing grease is not used to lubricate the fifth wheel.
Answer D is wrong. 85/140 axle lube is not used to lubricate the fifth wheel.

Question #94
Answer A is correct. A damaged gasket or missing sealant could cause a leak in this location.
Answer B is wrong. Overloading of the drive train would increase the axle temperature and thin out the fluid, but a good gasket should still seal this area.
Answer C is wrong. A plugged breather would increase the pressure inside the axle assembly, but this condition usually forces oil past the lips of the wheel seal, which is the weakest sealing point.
Answer D is wrong. Moisture contamination will have no effect on the ability of gaskets to seal or leak.

Question #95

Answer A is wrong. Power going to the right headlight would not be interrupted.
Answer B is wrong. Power going to the left headlight would not be interrupted.
Answer C is wrong. Power going to the right high beam would be unaffected.
Answer D is correct. The left high-beam headlight would be dim. Corrosion causes unwanted high circuit resistance, causing a decrease in current flow. Most electrical loads have a specific resistance designed into them to operate at a predicted voltage and current level. Any unwanted resistance in the circuit will affect the load device.

Question #96

Answer A is wrong. It is normal for drive axles to acquire slight condensation, which usually "boils off" during normal driving conditions.
Answer B is correct. Infrequent driving and short trips do not bring the axle to operating temperature, and moisture does not burn off.
Answer C is wrong. Although submerging the axles in water can introduce water into the axle, it is not the most likely cause.
Answer D is wrong. Any water that happens to make its way into the axle would normally be "boiled off" during normal driving conditions.

Question #97

Answer A is wrong. Mismatched or under-inflated tires will cause vehicle wandering.
Answer B is wrong. Springs matched with different spring rates will cause an imbalance, sagging, and/or loss of designed load-carrying ability, and could also possibly affect the designed riding geometry of the vehicle and thus cause wandering.
Answer C is wrong. Broken or missing spring leaves will cause a reduction in the load-carrying ability and of overall spring rate of the affected vehicle and will cause wandering.
Answer D is correct. A frame has to be within manufacturer's specifications.

Question #98

Answer A is correct. Record the observed problems on a PMI.
Answer B is wrong. Informing the service department will not log the observed problem.
Answer C is wrong. Servicing the bracket should not be done until the problem is logged on the PMI and authorization for the repair is obtained.
Answer D is wrong. Servicing the bracket should not be done until the problem is logged on the PMI and authorization for the repair is obtained.

Question #99

Answer A is wrong. Technician B is also correct.
Answer B is wrong. Technician A is also correct.
Answer C is correct. Both technicians are correct. The ½-inch measurement between the release bearing and clutch brake provides the necessary pedal travel to produce release bearing to clutch brake contact at the full pedal position. The ⅛-inch release fork to release-bearing clearance produces the correct clutch-pedal free play.
Answer D is wrong. Both technicians are correct.

Question #100

Answer A is wrong. The outboard S-cam bushings should be lubricated until grease appears at the inboard seal; if grease exits through the outboard seal it will end up in the drum assembly.
Answer B is wrong. Low pressure grease is sufficient to lubricate most chassis components and high pressure grease can cause seal damage and its use should be avoided except for lube operations such as spring pins.
Answer C is wrong. Neither technician is correct.
Answer D is correct. Neither technician is correct.

Question #101

Answer A is correct. The jaws are not reversible. The jaws should be inspected and replaced if worn. Answer B is wrong. The fifth wheel should be checked for horizontal movement during a PMI. When checking fifth-wheel adjustment, movement should not exceed ⅛-inch horizontal play at any time. Answer C is wrong. Fastener integrity and torque should be checked at each PMI.

Answer D is wrong. Locking mechanism should be inspected during a PMI. The locking mechanism may be released mechanically with a lever or by air pressure supplied to an air cylinder. Test the operation of either type during the PMI.

Question #102

Answer A is correct. Only Technician A is correct. The service brakes are tested during road tests of the vehicle. The brakes that are activated by the brake pedal are referred to as the service brakes. Work the brakes during the road test using a mixture of snub braking and aggressive panic stops. The FMVSS sets the standards used in the United States and Canada to define stopping distances as well as other braking safety criteria.

Answer B is wrong. A test drive is a mandatory procedure of a PMI.

Answer C is wrong. Only Technician A is correct.

Answer D is wrong. Only Technician A is correct.

Question #103

Answer A is correct. It is not necessary to replace the axle alignment washers when replacing a radius rod on a truck. If the washers are not worn beyond use, then they should be reused to ensure a correct realignment of the axle. Also, the number of washers and their respective positions (forward or rearward) should be noted and returned exactly as removed.

Answer B is wrong. Securing the vehicle is one of the first operations performed.

Answer C is wrong. Replacement rods come in many different sizes.

Answer D is wrong. An axle alignment check is always in order when replacing a torque rod.

Question #104

Answer A is wrong. Knock sensors are used only on gasoline engines.

Answer B is wrong. O_2 sensors are used only on gasoline engines.

Answer C is wrong. Only gasoline engines commonly use MAF.

Answer D is correct. Electronic fuel-injected diesel and gasoline engines use throttle position sensors. The TPS is a variable resistor that supplies the ECM with throttle position information so the ECM can regulate engine performance and is located in the electronic foot pedal assembly (EFPA). Many systems use a 5-volt reference voltage that is sent from the ECM to the TPS, and the TPS returns a portion of it, proportional to pedal mechanical travel. The actual voltage signal returned to the ECM is converted to counts, which can be read with an EST.

Question #105

Answer A is wrong. Technician B is also correct.

Answer B is wrong. Technician A is also correct.

Answer C is correct. Both technicians are correct. Idle speed can usually be raised and lowered by the use of a toggle switch provided on the dash on electronic engines. Idle shutdown duration is a programmable feature on most electronically controlled engines. Engine shutdown can be programmed to occur after a desired time interval. When the desired time has elapsed without input from the throttle pedal or clutch pedal, the engine will shut down automatically. To change this programmed time, an EST is needed.

Answer D is wrong. Both technicians are correct.

Question #106

Answer A is wrong. When the wiper blade on the driver's side is cracked and badly worn, the vehicle should be tagged out of service.

Answer B is wrong. When the wiper motor is inoperative, the vehicle should be tagged out of service.

Answer C is correct. When the passenger's side wipe blade is cracked and badly worn, the vehicle should not be tagged out of service.

Answer D is wrong. If the driver's side wiper arm is distorted so the blade does not contact the windshield properly, the vehicle should be tagged out of service.

Question #107

Answer A is wrong. A truck may not necessarily be pulled out of service if an exhaust leak does not meet the CVSA out-of-service criteria.

Answer B is correct. Only Technician B is correct. The vehicle can only be put out of service if the leaking exhaust fumes enter the driver/sleeper compartment or if they're likely to result in charring, burning, or damaging of the wiring, fuel supply or other combustible parts.

Answer C is wrong. Only Technician B is correct.

Answer D is wrong. Only Technician B is correct.

Question #108

Answer A is wrong. A simple visual inspection without having the engine assembly "moving" would be inaccurate since the mounts would not be under any dynamic stress.

Answer B is wrong. Correlating information from the driver and from the vehicle's maintenance history can only confirm a proper "dynamic" test of a truck's engine mounts.

Answer C is correct. The correct way to check engine mounts is to observe the engine and its mounts while the engine is being torqued with its brakes applied.

Answer D is wrong. A vehicle should never be deadlined because of a driver's suspicion.

Question #109

Answer A is wrong. This valve cannot be disassembled, cleaned, and reinstalled.

Answer B is correct. Only Technician B is correct. This air shift valve is not serviceable and must be replaced as a unit. The cost and time to disassemble many air valves, obtain parts and the light aluminum bodies make rebuilding impractical. Although some vehicles are equipped with electrical shift units, most axles are equipped with pneumatic shift systems. There are two air-activated shift systems predominantly used to select the range of a dual range tandem axle or to engage a differential lockout. Usually the air shift unit is not serviceable.

Answer C is wrong. Only Technician B is correct.

Answer D is wrong. Only Technician B is correct.

Question #110

Answer A is wrong. Technician B is also correct.

Answer B is wrong. Technician A is also correct.

Answer C is correct. Both technicians are correct. Maladjusted bearings fail rapidly, especially when preloaded. The result of high preload on a bearing can be to friction-weld it to the axle. All bearing manufacturers in North America have endorsed the Truck Maintenance Council (TMC) method of adjusting tapered roller wheel bearings. The bearing end play must be measured with a dial indicator. The required specification is between 0.001 inch and 0.005 inch. End play must be present. If at least 0.001 inch end play is not present, the adjustment procedure must be repeated.

Answer D is wrong. Both technicians are correct.

Question #111

Answer A is correct. Only Technician A is correct. The measurement should be between 1 and 2 inches with 1½ inches being typical. With the clutch pedal fully raised, there should always be ⅛ inch free-play between the release yoke fingers and the clutch release bearing pads. This free-play is taken up by the first 1 to 1½ inches of pedal travel. Measure the distance the pedal can be pushed down before thrust is exerted on the clutch release bearing to check clutch-pedal free travel.
Answer B is wrong. The manufacturer's recommendations should be checked on hydraulic clutches.
Answer C is wrong. Only Technician A is correct.
Answer D is wrong. Only Technician A is correct.

Question #112

Answer A is wrong. The applied pressure drop on a straight truck with no trailer should be no more than 4 psi in 2 minutes. Before performing a pressure-drop test, the system must be at full pressure. Make and maintain a brake application. Allow pressure to stabilize for one minute, and then begin timing for two minutes while watching the dash gauge.
Answer B is wrong. The technician holds the brake pedal down for the entire test to ensure a proper reading. A block of wood may be used to hold the brake pedal down during the pressure-drop test.
Answer C is wrong. Neither technician is correct.
Answer D is correct. Neither technician is correct.

Question #113

Answer A is wrong. The fuel inlet line is not part of an emissions systems check.
Answer B is correct. You should always check the PCV valve when any inspection of the emission control system is performed. The PCV system requires service at regular intervals to ensure that the system operates properly. Sludge and carbon can plug the PCV valve. Service the PCV valve by inspecting it, making sure that the valve is not clogged and it moves freely. Inspect all PCV system hoses for cracks and deterioration. Regular maintenance should include cleaning the PCV valve with a carburetor or fuel injector cleaner.
Answer C is wrong. The air cleaner vacuum hoses are not part of an emissions systems check.
Answer D is wrong. The intake manifold service port is not part of an emissions systems check.

Question #114

Answer A is wrong. SCA stands for supplemental coolant additive, which is an corrosion inhibitor additive. Adding more antifreeze does not change SCA reading.
Answer B is wrong. Continuing to run the truck until the next PMI will not change the SCA reading.
Answer C is correct. Drain the entire cooling system and add the proper SCA mixture to achieve the correct percentage of SCA.
Answer D is wrong. This will not change the SCA reading.

Question #115

Answer A is wrong. Technician B is also correct.
Answer B is wrong. Technician A is also correct.
Answer C is correct. Both technicians are correct. A proper inspection of a heavy-duty truck hydraulic braking system should include a visual of wheel cylinders. Look for evidence of leaks on the inboard side of the tire and wheel. Leakage and contamination around the wheel cylinders generally indicates leakage and that repairs are needed. Leaking wheel cylinders must be reconditioned or replaced. Brake lines should be in good condition with no corrosion or damage that may compromise system performance. Leaking or damaged brake lines must be replaced.
Answer D is wrong. Both technicians are correct.

Question #116

Answer A is correct. Only Technician A is correct. The engine/exhaust brake should be tested on a road test. Use the engine brake during the road test when the engine rpm is in torque rise. Note the disengage rpm.

Answer B is wrong. All engine compression brakes produce exhaust noise. When the compression brake is energized, it causes the engine to perform like a power-absorbing air compressor. The compression brake system uses engine oil pressure to open the exhaust valves before the compression stroke is complete and combustion does not occur in the cylinder. The compressed air is released into the engine exhaust system.

Answer C is wrong. Only Technician A is correct.

Answer D is wrong. Only Technician A is correct.

Question #117

Answer A is wrong. Side amber marker lights are a mandatory check. Check for burned-out, damaged, or missing marker lights and reflectors.

Answer B is wrong. All lighting, both interior and exterior, is to be inspected. Check for proper operation, color, and cleanliness of headlights, turn signals, stoplights, taillights, emergency flashers, reflectors, marker lights, clearance lights, cab instrumentation and interior lights, and trailer lighting.

Answer C is wrong. All wiring connectors are checked when performing an inspection. Wiring connectors should be clean and tight. Wires and/or harnesses must be replaced if insulation becomes burned, cracked, or deteriorated.

Answer D is correct. The reefer power supply is not part of the trailer's lighting system.

Question #118

Answer A is correct. The frame web may be drilled, but careful consideration should be given to the number, location, and sizes of additional frame bolt holes. Locate holes as close as possible to the horizontal center of the web, and do not place more than three bolt holes on the same vertical line in the web. Do not drill holes in heat-treated frames.

Answer B is wrong. Holes should never be drilled into the flange sections of the frame rails, as this weakens the frame.

Answer C is wrong. Cutting holes in the frame with a torch is not an accepted practice. Bolt holes should not be larger than the size of the bolt being installed.

Answer D is wrong. Notching the frame rails severely weakens the structure and could ultimately result in fracture and subsequent failure.

Question #119

Answer A is wrong. Technician B is also correct.

Answer B is wrong. Technician A is also correct.

Answer C is correct. Both technicians are correct. Air and electric horns are important safety devices enabling the operator to warn other motorists of imminent danger. Both are required to be fully functional. They should be tested during the driver's pre-trip and post-trip inspections. Verify their operations as part of each preventative maintenance inspection. Federal Motor Carrier Safety Regulations require a vehicle to have at least ½2 inch of tread depth at the most worn spot on steer axle tires. Measure the tread depth at the inboard, center, and outboard tread grooves, and at several places around the circumference of the tire.

Answer D is wrong. Both technicians are correct.

Question #120

Answer A is wrong. The low system vacuum does not necessarily condemn a vehicle to be out of service, but will require a thorough diagnostic inspection by a certified technician depending upon the severity of the loss.

Answer B is wrong. An engine with low performance can have many causes, not exclusively low vacuum. Outside of direct leakage to the atmosphere, low vacuum is generally a result of low engine performance, not a cause of it.

Answer C is correct. The vacuum reserve is insufficient to permit one full brake application after the engine is shut off.

Answer D is wrong. There are no set system-wide vacuum bleed down standards.

Question #121
Answer A is wrong. Channeling is not a breaking down of a lubricant.
Answer B is wrong. Extreme pressures have nothing to do with lubricant separation.
Answer C is wrong. Flowing liquids always seek their lowest center of gravity, regardless of location.
Answer D is correct. Cold-temperature operation can cause transmission and axle lubrication problems known as channeling. High-viscosity oils typically used in transmissions and axles thicken causing greatly reduced flow, or they may not flow at all when the vehicle is first started. Channeling leads to components not getting proper lubrication. Rotating gears can push lubricant aside leaving voids or channels where no lubricant is actually touching the gears.

Question #122
Answer A is wrong. The inspection should not start on the right side of the vehicle moving in a clockwise direction.
Answer B is correct. A circle-check or pre-trip inspection begins on the left front side of the truck and continues in a counterclockwise direction around it. An adaptation of the Commercial Vehicle Safety Alliance (CVSA) would use the following step process: vehicle overview, engine compartment, inside the cab (drivers side), front of cab, left side of cab, left fuel tank area, left rear tractor area, left side of trailer, rear of trailer, right side of trailer, right rear tractor area, right fuel tank area, right side of cab, cab.
Answer C is wrong. The inspection should go from the left front to back in a counterclockwise direction, not side to side.
Answer D is wrong. The inspection should not start at the rear and go to the front.

Question #123
Answer A is wrong. Technician B is also correct.
Answer B is wrong. Technician A is also correct.
Answer C is correct. Both technicians are correct. During a normal PMI, an engine oil sample is taken for inspection of its overall lubricating qualities, plus metal and dirt content. This oil may also be tested for corrosion-causing agents such as sulfuric and hydrochloric acids, and some esters. It is for this reason that the engine oil should be at normal operating temperature to ensure that all possible contaminants and corrosives be contained in the oil in its normal operating condition. The oil sample should be taken during an oil change at the midpoint of the drainage. The test bottle should be clean before the sample, and sealed and labeled afterwards.
Answer D is wrong. Both technicians are correct.

Question #124
Answer A is wrong. A shorted armature winding would cause increased current flow in the insulated side of the circuit.
Answer B is wrong. An open field coil would stop or interrupt current flow on the insulated side.
Answer C is correct. A damaged or corroded battery ground cable could show a voltage drop of more than 0.5 volts. Many cranking motors and batteries are needlessly replaced due to defective or corroded cables and terminals. A voltage drop value exceeding 0.2 volt through any individual terminal or cable indicates a possible problem. Electrical connections should not produce voltage drops greater than 0.1 volt and total voltage drop of all the cables and terminals in a cranking circuit should not exceed 0.5 volt.
Answer D is wrong. The ignition-switch circuit is not related to the ground-circuit resistance test.

Question #125
Answer A is wrong. A pour point check of the fuel is not done on a PMI.
Answer B is wrong. Dash gauge checks are part of an electrical system check.
Answer C is wrong. A test of the pump is only performed when there is a performance problem.
Answer D is correct. Replacing the fuel filter and priming the system is done during a PMI. Filters must be primed with fuel before the engine can be started. Most fuel filters can be filled with clean fuel and reinstalled, but some manufacturers recommend that the secondary filter be installed dry and primed with a hand pump. This procedure ensures that only filtered diesel fuel enters the secondary filter. Fuel contamination can occur easily if the filter is primed with unfiltered fuel or dirt is allowed to enter when the element is being changed.

Question #126

Answer A is wrong. Measuring from the road surface to the lower plate would be inaccurate.

Answer B is correct. A cab air suspension system is shown in the figure. When checking cab ride height, the truck should be on a level surface and the cab should appear to be level. Make sure system air pressure is at specified pressure before performing measurements. Use a tape measure to check cab ride height at the location specified by the manufacturer. This is usually between the upper and lower plates. Compare readings with those specified by the manufacturer and adjust the leveling valve as necessary.

Answer C is wrong. Measuring from the road surface to the upper plate would be inaccurate.

Answer D is wrong. Measuring the angle of the regulator valve arm would be inaccurate.

Question #127

Answer A is wrong. Technician B is also correct.

Answer B is wrong. Technician A is also correct.

Answer C is correct. Both technicians are correct. A plugged breather can cause a pressure build-up in the axle, forcing fluid past the seals and gaskets. Carefully investigate any source leaks found on the axle differential. Replacing the seal if the axle differential breather is plugged will not cure leaking seals and the problem will continue to exist. Grease fittings should be cleaned off before lubricating to remove accumulated grease and abrasives. Otherwise, dirt and contaminants will be forced into the component being lubricated. While supplying grease through the grease fittings, check to be sure they are not missing, damaged, broken, or plugged.

Answer D is wrong. Both technicians are correct.

Question #128

Answer A is wrong. Technician B is also correct.

Answer B is wrong. Technician A is also correct.

Answer C is correct. Both technicians are correct. Technicians performing refrigeration service must be certified to do so. A manifold gauge set is used to recycle, recharge, and diagnose A/C systems. The manifold gauge set is designed to access the refrigeration circuit and control refrigerant flow when adding or removing refrigerant. When the manifold test set is connected into the system, pressure is registered on both of the gauges at all times. Make certain that both hand valves on the manifold gauge set are closed before connecting to the system to avoid releasing refrigerant or contaminating the system.

Answer D is wrong. Both technicians are correct.

Question #129

Answer A is correct. Only Technician A is correct. Hub-piloted wheels are less susceptible to loosening and cracking than stud-piloted wheels. The hub-piloted system simplifies centering and clamping wheels to the hubs because the hub centers the wheel and the nuts and studs provide only clamping force. The hub-piloted wheel uses one cap nut and eliminates the need for inner cap nuts.

Answer B is wrong. Generally, spoke wheels experience greater alignment and balance problems than disc designs. Proper installation procedures are critical on spoke wheels. If the rim clamps are not properly installed, wheel out-of-round or wobble may be excessive.

Answer C is wrong. Only Technician A is correct.

Answer D is wrong. Only Technician A is correct.

Question #130

Answer A is wrong. The pull circuit allows the cab to come back over the engine.

Answer B is wrong. The pull circuit allows the cab to come to the 45 degree tilt position.

Answer C is wrong. The pull circuit allows the cab to descend to the 60 degree overhead position.

Answer D is correct. The push circuit raises the cab from the lowered position to the desired tilt position. Remember that in most systems, whenever raising or lowering the cab, stop working the hydraulic pump once the cab goes over center. The cab falls at a controlled rate and continued pumping could lock up the tilt cylinder. Inspect the hydraulic cylinders for external leakage and determine if the cab can be raised and lowered. Check for internal leakage by observing bleed down. Make sure the mechanical stops engage and hold safely.

Question #131

Answer A is wrong. A carbon pile and ammeter gauge is the required equipment for testing the output of a battery, not what is shown.

Answer B is wrong. An ohmmeter connected in series with the ignition switch is the required equipment and operating procedure.

Answer C is wrong. A voltmeter connected to the ground side of the starter circuit is required for a voltage drop test on the starter ground side.

Answer D is correct. The technician is checking the starter relay operation. The starter relay is a remote mounted magnetic switch that uses low current to control high current flow from the battery to the starter motor. If the starter system has a magnetic switch, jumper around the heavy terminals to see if the motor cranks. If the starter motor cranks, the relay is defective, assuming that control current from the starting switch is available at the small terminal of the relay.

Question #132

Answer A is correct. A damaged or failed vibration damper can cause crankshaft failure. The primary vibration damper function is to reduce crankshaft torsional vibration. The damper housing is coupled to the crankshaft and uses springs, rubber, or viscous fluid to drive the inertia ring. Visually inspect the damper housing, noting any dents or signs of warpage. Check for indications of fluid leakage with damper in place. Trace evidence of leakage justifies replacement of the damper. With the damper on the crankshaft, check for wobble using a dial indicator. Viscous dampers should be checked for nicks, cracks, or bulges as well.

Answer B is wrong. A visible dent in the vibration damper housing will not cause engine misfire.

Answer C is wrong. The most serious consequence would be engine crankshaft failure.

Answer D is wrong. The most serious consequence would be engine crankshaft failure.

Question #133

Answer A is wrong. Retorquing of wheel lug nuts is required when performing a PMI. At recommended intervals retorquing is necessary. This is due to the constant flexing of the stud/hub material. Overloading can be a factor. Manufacturers have specified intervals recommended for wheel and rim inspections.

Answer B is wrong. Increased load-carrying capacity is not an advantage of switching to low-profile tires. Low-profile tires decrease the load-carrying capacity of a vehicle. The low-profile tire does offer several advantages compared to conventional tires, which include reduced rolling resistance, cooler running, longer tread life, improved fuel economy, reduced vehicle height with a lower center of gravity, and improved traction.

Answer C is wrong. Neither technician is correct.

Answer D is correct. Neither technician is correct.

Question #134

Answer A is wrong. Technician B is also correct.

Answer B is wrong. Technician A is also correct.

Answer C is correct. Both technicians are correct. Begin battery service with a visual inspection looking for loose hold-downs, defective cables, damaged terminal posts, loose terminal connections, clogged vents, corrosion, dirt or moisture, cracked case, and battery box condition. Following a visual inspection, a battery open circuit voltage test can be used to determine its state of charge. A battery that has recently been charged by running the engine or using a battery charger must have its surface charge removed before performing the open circuit voltage test. A stabilized open circuit voltage reading of 12.6 or more indicates a fully charged battery.

Answer D is wrong. Both technicians are correct.

Question #135

Answer A is correct. Axle lubrication level is correct only when the fluid level is even with the bottom of the filler hole in the carrier housing with the truck on level ground. The angle of the drive pinion, as mounted under the vehicle, determines which oil fill/level plug is to be used if two are present. Inadequate lubrication from low lube levels could create friction, cause overheating, break down the protective film, and finally result in seizure or welding of mating components. The other choices could indicate a low fluid level.
Answer B is wrong. Lube level must be even with the bottom of the filler hole.
Answer C is wrong. Lube level must be even with the bottom of the filler hole.
Answer D is wrong. Lube level must be even with the bottom of the filler hole.

Question #136

Answer A is wrong. CVSA out-of-service criteria indicate cracks of ¼ inch in length or more.
Answer B is correct. Only Technician B is correct. Check the windshield glass for cracks, dirt, illegal stickers, or discoloration that obstruct the driver's view. Any crack over ¼-inch wide, intersecting cracks, discoloration not applied in manufacture, or other vision-distorting matter in the sweep of the wiper on the driver's side of the windshield will take a vehicle out of service by the commercial CVSA out-of-service standards.
Answer C is wrong. Only Technician B is correct.
Answer D is wrong. Only Technician B is correct.

Question #137

Answer A is wrong. VORAD stands for Vehicle On-board Radar.
Answer B is wrong. Anti-collision devices are not mandatory. Because anti-collision protection is high on the priority list of the Federal Highway Administration, collision warning systems will most likely become mandatory on heavy-duty vehicles in the near future.
Answer C is wrong. Neither technician is correct.
Answer D is correct. Neither technician is correct. VORAD (Vehicle On-Board Radar) is a collision warning system (CWS) that uses radar technology to sense some specific vehicle danger conditions. VORAD is used in trucks to sense straight-ahead closing velocities and proximity as well as blind-side proximity. The system is electronically managed and consists of relatively few components. It must be connected into the vehicle electrical and electronic system so it can share sensor signals such as the input from the vehicle road speed sensor

Question #138

Answer A is correct. Only Technician A is correct. A clutch brake needs to be inspected for wear and fatigue. Worn clutch brake friction surfaces can lead to clashing (grinding) when moving the gear lever into reverse or low gear while the vehicle is stationary.
Answer B is wrong. The friction face needs to be inspected. The clutch brake is a circular disc with a friction surface that is mounted on the transmission input spline shaft using two tangs that engage machined slots. The clutch brake is mounted between the release bearing and the transmission. Its purpose is to slow or stop the transmission input shaft from rotating in order to allow gears to be engaged without clashing and to keep transmission gear damage to a minimum. Clutch brakes are used only on vehicles with nonsynchronized transmission.
Answer C is wrong. Only Technician A is correct.
Answer D is wrong. Only Technician A is correct.

Question #139

Answer A is correct. Only Technician A is correct. While checking the cooling system, checking the records of coolant usage is important to indicate whether excess coolant is being lost. A proper inspection of the cooling system should include a check of the cooling system maintenance history, coolant change intervals, component replacement(s), and past diagnoses.

Answer B is wrong. The coolant is changed too frequently and does not take into consideration factors such as whether extended-life coolant is being used, the time interval between the last cooling system drain and refill, or what scheduled PMI is being performed. Identify the coolant being used and be sure to follow the pertinent change intervals and refill with the correct type. ELCs have a longer service life than ethylene glycol (EG) and propylene glycol (PG) coolants.

Answer C is wrong. Only Technician A is correct.

Answer D is wrong. Only Technician A is correct.

Question #140

Answer A is correct. Only Technician A is correct. You always lubricate the slack adjusters whenever an inspection of the brakes is scheduled. Slack adjusters are lubricated by grease or automatic lubing systems. The seals in slack adjusters are always installed with the lip angle facing outward. When grease is pumped into the slack adjuster, grease exits easily past the seal lip when the internal lubrication circuit has been charged.

Answer B is wrong. All brake chamber pushrods must be checked during a brake inspection. Air brake chambers should be checked for signs of damage. Visually check for damage from road hazards and look for cracks around mounting studs. Check the air chamber bracket for cracks, looseness, or audible air leaks at brake chamber that might indicate a ruptured diaphragm, loose chamber clamp, or other failure.

Answer C is wrong. Only Technician A is correct.

Answer D is wrong. Only Technician A is correct.

Question #141

Answer A is wrong. Replacing the breather is not cleaning the breather.

Answer B is wrong. Soaking the breather in gasoline is not a recommended procedure. Gasoline is not an acceptable solvent because of its extreme combustibility. It is unsafe in the workshop environment. There are many suitable commercial solvents available that meet current environmental and fire safety standards.

Answer C is correct. The best way to clean a rear-axle breather is to soak in solvent and then use compressed air. The rear-axle breather should be cleaned each time the lube level is checked. Investigate any source of leaks found on the axle differential. Replacing a seal if the axle differential breather is plugged does not cure leaking seals. A plugged breather produces high-axle housing pressure, which can result in leakage past seals or gaskets.

Answer D is wrong. Wiping the orifice will not clean any dirt trapped inside the breather.

Question #142

Answer A is wrong. If a tire is mounted or inflated so it contacts any part of the vehicle, this problem must be corrected before placing the vehicle in service.

Answer B is wrong. If a drive axle tire has a tread depth of less than $\frac{1}{32}$ inch at two adjacent tread grooves at three separate locations, this problem must be corrected before placing the vehicle in service.

Answer C is correct. If a steer axle tire has a tread depth of $\frac{5}{32}$ inch at two adjacent tread grooves at any location on the tire, the tire is satisfactory. Minimum tread depth is $\frac{4}{32}$ inch.

Answer D is wrong. If the tire is labeled "Not for Highway Use," this problem must be corrected before placing the vehicle in service.

Question #143

Answer A is wrong. Checking the placement of the actuator arm is part of a valid inspection. A self-adjusting clutch that is out of adjustment could be caused by an actuator arm incorrectly inserted into the release bearing sleeve retainer. Visually check to see that the actuator arm is correctly inserted. If the assembly is properly installed, the spring will move back and forth as the pedal is full stroked. The clutch will not compensate for wear if the actuator arm is not inserted into the release sleeve retainer, or if the release bearing travel is less than ½ inch.

Answer B is wrong. Checking for a bent adjuster arm is a valid inspection. A self-adjusting clutch found out of adjustment could be caused by a bent adjuster arm.

Answer C is wrong. Checking for a frozen adjusting ring is a valid inspection. The adjusting ring should be checked for damage or a frozen condition.

Answer D is correct. Clutch shaft runout is not checked on a self-adjusting clutch that is out of adjustment.

Question #144

Answer A is correct. Only Technician A is correct. Wear between the clevis pin and yoke causes mechanical slop, which results in overstroking.

Answer B is wrong. Properly functioning automatic slack adjusters should not require frequent adjustment. Manually adjusting automatic slack adjusters is dangerous and should not be done, except during installation or in an emergency to move the vehicle to a repair facility, because manual adjustment of this brake component (1) fails to address the true reason why the brakes are not maintaining adjustment, giving the operator a false sense of security about the effectiveness of the brakes, which are likely to go out of adjustment again soon, and (2) causes abnormal wear to the internal adjusting mechanism for most automatic slack adjusters, which may lead to failure of this brake component. (H-06-11).

Answer C is wrong. Only Technician A is correct.

Answer D is wrong. Only Technician A is correct.

Question #145

Answer A is correct. When the transmission breather is plugged, internal pressure can build inside the transmission and push oil out seals and gaskets. Always check the transmission breather filters when repairing any transmission leaks. Under normal operating conditions, the transmission can build up pressure inside the case if the vents or breathers are not functioning correctly. This pressure can cause fluid to leak past seals that do not need replacement or seals that are otherwise in good working order.

Answer B is wrong. The breather could be plugged. The most common cause of gaskets appearing to be blown out of the shift housing cover would be a plugged breather.

Answer C is wrong. The breather could be plugged. The most common cause of gaskets appearing to be blown out of the shift tower would be a plugged breather.

Answer D is wrong. The breather could be plugged. When a breather becomes plugged, fluid is often forced past seals in the transmission.

Question #146

Answer A is wrong. Technician B is also correct.

Answer B is wrong. Technician A is also correct.

Answer C is correct. Both technicians are correct. A faulty fan clutch can cause above-normal engine temperatures or overheating. Fan clutch types may be viscous, thermatic, air apply, air release, electric, and hydraulic. Each type has different characteristics and should first be identified to avoid improper diagnosis. Most pneumatic fan clutch failures are caused by air leaks. Viscous and hydraulic units should not exhibit any signs of fluid leakage. Radiators, because of their location, are susceptible to damage and restriction. Buildup of road debris and summer bugs can severely compromise the radiator's ability to effectively cool. Radiators should be cleaned externally using either a low-pressure steamer or regular hose, detergent, and a soft nylon bristle brush. Never use a high-pressure washer because this will almost certainly result in damaging the cooling fins.

Answer D is wrong. Both technicians are correct.

Question #147

Answer A is wrong. The most common method to link the tractor's light system to the trailer's and power various lights on a trailer requires the use of a seven-way receptacle and trailer cord plug mated together. Many vehicle manufactures use colored wires to more easily identify common electrical circuits.

Answer B is wrong. A circle-check or pre-trip inspection begins on the left front side of the truck and continues in a counterclockwise direction around it. An adaptation of the CVSA would use the following step process: vehicle overview, engine compartment, inside the cab (drivers side), front of cab, left side of cab, left fuel tank area, left rear tractor area, left side of trailer, rear of trailer, right side of trailer, right rear tractor area, right fuel tank area, right side of cab, cab.

Answer C is wrong. Neither technician is correct.

Answer D is correct. Neither technician is correct.

Question #148

Answer A is correct. Only Technician A is correct. When considering whether to deadline a truck, the importance of moving a load must be weighed against the probability that a mechanical condition will result in a breakdown. The Federal Motor Carrier Safety Regulations, Part 393, say that anything mechanical that can either cause or prevent an accident is a safety item. All alleged safety problems should be addressed. The regulations also forbid dispatching a vehicle that is likely to break down. Regardless of whether any fine might be levied for that, dispatching a truck that is on the brink of breaking down is not just good business, nor is it safe.

Answer B is wrong. Items such as a malfunctioning cruise control or an engine brake are not deadline items and would not warrant deadlining the truck.

Answer C is wrong. Only Technician A is correct.

Answer D is wrong. Only Technician A is correct.

Question #149

Answer A is wrong. Technician B is also correct.

Answer B is wrong. Technician A is also correct.

Answer C is correct. Both technicians are correct. Coolant hoses should be checked for cracks, hardness, deterioration, chafing, and bulges every time a general inspection is performed. All hose clamps should be checked for tightness and/or deterioration. Check for leaks around hoses and clamps both when cold and at operating temperature. A general coolant system inspection should be made every time a vehicle is scheduled for a PMI. Inspect the radiator for airflow restriction, leaks, damage, and mountings. Check operation of the fan clutch and inspect the fan assembly and shroud. Pressure test the cooling system and radiator cap. Inspect the coolant recovery system and water pump. Check the coolant for contamination, supplemental coolant additives (SCAs), and protection level. Service coolant filter/conditioner and drain/refill cooling system if scheduled.

Answer D is wrong. Both technicians are correct.

Question #150

Answer A is wrong. Technician B is also correct.

Answer B is wrong. Technician A is also correct.

Answer C is correct. Both technicians are correct. Linings with a thickness less than ¼ inch at the shoe center for air drum brakes should be replaced. Brake linings that are saturated with oil, grease, or brake fluid should be replaced. Linings and pads should be firmly attached to the shoe. Unequal lining wear should be noted and the cause investigated. When replacing linings, both wheels of an axle should be replaced to maintain brake balance. The frictional rating of linings is coded by letter codes stamped on the edge of the lining. Combination sets of shoe linings are occasionally used, and these linings use different friction ratings on the primary and secondary shoes. When combination friction lining sets are used, care should be taken to install the lining blocks in the correct locations on the brake shoes.

Answer D is wrong. Both technicians are correct.

Question #151

Answer A is wrong. When inspecting the driveline, the transmission output shaft and rear-axle input shaft should be checked because they can affect the shaft's harmonics. End yokes should be checked for retaining nut looseness, radial looseness, fractures, runout, and vertical alignment, which could cause vibration, premature wear of powertrain components, and failure of U-joints.

Answer B is correct. Only Technician B is correct. A technician should always check the trunnions, tubes, and center support bearing for damage. Common U-joint damage includes cracking of trunnions or spider, galling, spalling, pitting, and brinelling. Check the slip spline for excessive radial movement. Check the shaft for damage, bent tubing, missing balance weights, or foreign material on the shaft. The center bearing should be checked for leaking lubricant, damaged seals, and proper mounting.

Answer C is wrong. Only Technician B is correct.

Answer D is wrong. Only Technician B is correct.

Question #152

Answer A is wrong. The valve on the left is the trailer air supply valve. It is octagonal and red in color. The trailer air supply valve is used to supply air to apply and release the trailer parking brakes. The valve is pulled outward to vent air pressure in the trailer parking chambers and to allow the chamber springs to apply the parking brakes. When the trailer air supply valve is pushed inward, air pressure is supplied through the trailer supply line to the trailer parking brake chambers to release the parking brakes.

Answer B is correct. Only Technician B is correct. The valve on the right in the figure is the system park valve. It is diamond shaped and yellow in color. The park control (system park) valve is used to apply and release the tractor and trailer parking brakes. If the park control valve is pushed inward, the parking brakes are released, and pulling the park control valve outward results in parking brake application.

Answer C is wrong. Only Technician B is correct.

Answer D is wrong. Only Technician B is correct.

Question #153

Answer A is wrong. It is not possible to "bounce" a heavy truck up and down to test its shocks. Even if it bounced, it would not be a good indicator of shock wear.

Answer B is correct. A technician should give the truck a thorough visual inspection of all suspension components for signs of fatigue, deterioration, and/or failure. On leaf-spring suspensions, check spring U-bolts, spring pins and bushings, shackles, spring brackets, hangers, and spring assemblies. Springs should be inspected for cracks, sag, and surface wear. Multiple leaf spring assemblies should be replaced as an assembly. On air suspension systems, check air springs, mounts, hoses, valves, linkage, and fittings for leaks or damage.

Answer C is wrong. A road test will not always give accurate results of the condition of a suspension or even steering linkage.

Answer D is wrong. A check of a PM schedule for repairs should only be made after a thorough inspection on the status of the vehicle.

Question #154

Answer A is wrong. Mounting tires of different construction on the same axle is mismatching. Radial-ply and bias-ply tires must not be mixed on the same axle. All the tires on an axle must be either radial ply or bias ply. On trucks with two or more drive axles, the tire matching on all the axles are necessary so all the tires are the same type. Because radial tires defect more than bias-ply tires under load, a mix of bias-ply and radial-ply tires overloads the bias-ply tires. This action causes premature tire tread wear and possible tire blowout.

Answer B is correct. Though it is not desirable to mix tread designs, both tires are the same nominal dimension, so this is least likely.

Answer C is wrong. Installation of a new tire next to an old tire is considered mismatching.

Answer D is wrong. Mounting tires of different circumferences and diameters is always considered mismatching. Matching tire sizes on dual wheels prevents tire tread wear from slippage due to uneven surface areas. Dual drive tires may be measured with a tape measure, string gauge, square, straightedge, or caliper.

Question #155

Answer A is wrong. It is not good practice to replace parts before finding the fault.

Answer B is wrong. Most sensors need at least 5 volts to operate properly. Some instrument panel gauges require protection against heavy voltage fluctuations that could damage the gauges or cause them to give incorrect readings. A voltage limiter provides this protection by limiting voltage to the gauges to a preset value. The limiter contains a heating coil, a bimetal arm, and a set of contacts. When the ignition is in the on or accessory position, the heating coil heats the bimetal arm, causing it to bend and open the contacts. This action cuts the voltage from both the heating coil and the circuit. When the arm cools down to the point where the contacts close, the cycle repeats itself. Rapid opening and closing of the contacts produces a pulsating voltage at the output terminal that averages 5 volts. Voltage limiting may also be performed electronically.

Answer C is wrong. Neither technician is correct.

Answer D is correct. Neither technician is correct.

Question #156

Answer A is correct. Only Technician A is correct. It is necessary to ensure that air is bled from a cooling system when refilling the system. This can be done by opening a hose or fitting on the top of the engine before the thermostat and filling the system until liquid coolant appears. This prevents trapped air from creating hot spots when the engine is first started. Some systems employ bleed valves to aid in removing trapped air from the cooling system.

Answer B is wrong. ELC coolants do not require in-service SCA conditioner monitoring. No inhibitor test kits are required to monitor the ELC SCA level. If a coolant filter is used, replace the existing filter with an SCA-free filter. Extended-life coolants promise a service life of 600,000 miles or six years with one additive recharge at 300,000 miles or three years. This compares with a typical service life of two years during which up to twenty recharges of SCA would be required for conventional EG and PGs.

Answer C is wrong. Only Technician A is correct.

Answer D is wrong. Only Technician A is correct.

Question #157

Answer A is wrong. Technician B is also correct.

Answer B is wrong. Technician A is also correct.

Answer C is correct. Both technicians are correct. You do block the wheels before attempting to adjust service brakes. The air reservoir pressure must be dropped below cut-in pressure before effective cycle resumes. Air brake system pressure in most trucks is set at a value between 110 and 130 psi (758 and 896 kPa), with 120 psi (827 kPa) being typical. System pressure is known as cut-out pressure, the pressure at which the governor outputs the unloader signal to the compressor. The unloader signal is maintained until pressure in the supply tank drops to the cut-in value. Cut-in pressure is required by FMVSS 121 to be no more than 25 psi (172 kPa) less than the cut-out value. The difference on most systems ranges between 20 and 25 psi (137 and 172 kPa). To check air governor cut-in pressure, with the engine running, build system pressure to achieve cut-out. Next, pump the service brake foot valve to drop pressure. Listen for the moment the compressor resumes loaded cycle; this is governor cut-in.

Answer D is wrong. Both technicians are correct.

Question #158
Answer A is wrong. Technician B is also correct.
Answer B is wrong. Technician A is also correct.
Answer C is correct. Both technicians are correct. Both the supply tank and the primary axle service reservoir must be drained before beginning the test. The air system should be fully charged (minimum 100 psi), and the spring parking brakes must be released. Next, locate and drain the supply tank and then the primary axle service reservoir completely by means of the draincocks. The primary gauge needle will drop to zero psi, indicating a pressure loss for that system, which is normal. The secondary gauge needle must remain at system pressure (minimum 100 psi) to start the check of the inversion valve. Depress the foot valve in a normal manner. The spring brakes on the rear axle(s) will apply as the air is exhausted from the brake chambers. Release the foot valve, and the spring brakes will release as air is supplied to them from the secondary reservoir. If the spring brakes do not or only partially release, then the inversion valve has failed and should be replaced. An inversion valve is a failsafe device and does not operate when brakes are functioning properly.
Answer D is wrong. Both technicians are correct.

Question #159
Answer A is wrong. No leakage or wetness of fuel is permitted on or around fuel tanks, mountings, fuel lines, or caps. When fuel is drawn out of a tank, it is replaced by air. This air always contains some percentage of moisture and in areas of high humidity, this can be extreme. Cooling temperatures cause this vaporized moisture to condense as water, which settles at the bottom of the tank. It is good practice to keep fuel tanks full, especially after the vehicle has been run and the tanks and fuel are warm. This keeps moisture-laden air out of the tanks and reduces water-in-fuel problems.
Answer B is wrong. The time of season has no bearing on draining water from fuel tanks.
Answer C is wrong. Neither technician is correct.
Answer D is correct. Neither technician is correct. The fuel tank mounts should be securely attached to the frame and the tank free of any cracks, corrosion, and damage that could cause future failure. Check to ensure that the fuel tank cap seal is in place and in good condition. The fuel lines, including the fuel crossover line must be leak free and should be secured by loom or brackets. Fuel lines should be checked for chafing and the possibility of future leaks or breakdowns.

Question #160
Answer A is wrong. Checking the road speed limiter is part of a vehicle road test. Connect the appropriate engine scan tool, OEM special tool, or laptop computer to the truck's diagnostic connector. During the road test, activate the tool's parameter specification readout and record the road speed. Reprogram if required.
Answer B is correct. A snapshot test for active codes is a diagnostic procedure that would not be performed as part of a normal scheduled PMI. A snapshot test readout from the ECM can be used to facilitate troubleshooting intermittent problems that either do not generate codes or log a code with no clear reason.
Answer C is wrong. Observing exhaust for excessive smoke is part of a road test. The color and consistency of the exhaust can reveal combustion problems. Oil entering the combustion chambers causes blue smoke in the exhaust. Unburned fuel, insufficient intake air, or an improper grade of fuel produces a gray or black smoke. Water or coolant entering the combustion chambers produces a white steam in the exhaust. White smoke may also indicate that injectors are misfiring. White smoke also occurs either when the engine starts the first time or when the ambient temperature is too low for proper combustion.
Answer D is wrong. Verification of proper engine brake operation is performed during a road test. Use the engine brake during the road test when the engine rpm is in torque rise. Note the disengage rpm.

Question #161

Answer A is wrong. A defective air governor will not cause a slow build-up, only cut-in and cut-off problems. The governor manages system pressure. It monitors pressure in the supply tank by means of a line directly to it. The governor manages compressor loaded and unloaded cycles. Loaded cycle is the compressor effective cycle; that is, the compressor is pumping air. The unloaded cycle occurs when the compressor is driven by the engine but it is not actually compressing air. The compressor is in the loaded cycle until is receives an air signal from the governor to put it into the unloaded cycle. The governor simply controls whether the compressor is in the loaded or unloaded operating mode. Air brake system pressure (cut-out pressure) in most trucks is set at values between 110 and 130 psi (758 and 896 kPa) with 120 psi (827 kPa) being typical. Cut-in pressure is required by FMVSS 121 to be no more than 25 psi (172 kPa) less than the cut-out value.

Answer B is wrong. A leak in the service air brake chamber will not cause slow build-up because it is on the application side, not the supply side.

Answer C is wrong. An air leak in the cab will not cause slow build-up since the air supply is protected by pressure-protection valves. A pressure-protection valve traps air in the air brake system if a leak occurs in one of the air-operated devices on the truck. When the pressure in air-operated devices are equal to the reservoir pressure, the pressure-protection valve remains open. If an air leak occurs in one of the air-operated devices, the pressure in the device drops below reservoir pressure. Under this condition, the pressure protection valve closes to prevent any further escape of air.

Answer D is correct. A defective air compressor is the most likely cause of slow build-up.

Answers to the Test Questions for the Additional Test Questions Section 6

1. D	22. A	43. D	63. B
2. A	23. D	44. C	64. C
3. A	24. A	45. A	65. C
4. B	25. C	46. C	66. B
5. A	26. B	47. D	67. A
6. A	27. D	48. B	68. D
7. A	28. D	49. D	69. C
8. A	29. A	50. A	70. D
9. D	30. C	51. B	71. B
10. C	31. C	52. B	72. D
11. C	32. C	53. B	73. A
12. C	33. C	54. D	74. A
13. C	34. D	55. D	75. C
14. C	35. D	56. A	76. D
15. B	36. B	57. A	77. B
16. A	37. B	58. B	78. D
17. C	38. C	59. C	79. C
18. A	39. C	60. C	80. C
19. D	40. C	61. B	81. D
20. B	41. D	62. C	82. C
21. C	42. B		

Explanations to the Answers for the Additional Test Questions Section 6

Question #1
Answer A is wrong. A water pump pulley would be above the vibration damper, not to the side of it.
Answer B is wrong. An air pump is not part of a diesel engine assembly.
Answer C is wrong. The fuel pump is not belt-driven.
Answer D is correct. An idler pulley is shown in the figure.

Question #2
Answer A is correct. The bulkhead is a trailer body component.
Answer B is wrong. Tires are a trailer underside component.
Answer C is wrong. The suspension system is a trailer underside component.
Answer D is wrong. The landing gear is a trailer underside component.

Question #3
Answer A is correct. Parallel connections are shown in the figure.
Answer B is wrong. Parallel-series connections would not be connected in this manner.
Answer C is wrong. Series would have all the connections on both sides connected this way.
Answer D is wrong. Series-parallel connections would have the ground side circuit in series.

Question #4
Answer A is wrong. The hub bearings generally do not fail prematurely.
Answer B is correct. Most failures are caused by air leaks.
Answer C is wrong. An electrical problem will cause the clutch to shut down, a failsafe device.
Answer D is wrong. Obstructions generally cause immediate damage to the fan blades.

Question #5
Answer A is correct. Identification lights are typically roof-mounted in this spot.
Answer B is wrong. Roof-mounted solar panels are not commonly used on trucks.
Answer C is wrong. Skylights are not generally used on a heavy truck.
Answer D is wrong. Air vents are not as common on the roof as clearance lights.

Question #6
Answer A is correct. The figure shows unequal shoe lining wear.
Answer B is wrong. Poor contact at center would have lining ends worn excessively.
Answer C is wrong. This would appear to gradually decrease in thickness from one end to the other.
Answer D is wrong. This would indicate a fairly uniform wear appearance.

Question #7
Answer A is correct. Zerk fittings are NOT part of the automatic chassis lubrication system.
Answer B is wrong. The reservoir is part of the automatic chassis lubrication system.
Answer C is wrong. The pump is part of the automatic chassis lubrication system.
Answer D is wrong. Distribution lines are part of an automatic chassis lubrication system.

Question #8
Answer A is correct. An incorrect ratio will only affect the truck's mileage and pulling ability, not cause failure.
Answer B is wrong. Normal wear always eventually causes a failure.
Answer C is wrong. Incorrect or lack of lubrication will always cause a failure.
Answer D is wrong. Fatigue will cause a drive axle failure.

Question #9
Answer A is wrong. Technician certification is not covered under the Clean Air Act.
Answer B is wrong. Standard service and PMI records cover this area.
Answer C is wrong. This is part of a PMI.
Answer D is correct. Refrigerants and their handling are mandated by the Federal Clean Air Act Amendment.

Question #10
Answer A is wrong. Technician B is also correct.
Answer B is wrong. Technician A is also correct.
Answer C is correct. Both technicians are correct. Active trouble codes cannot be erased. The event audit trail codes can be read.
Answer D is wrong. Both technicians are correct.

Question #11
Answer A is wrong. Technician B is also correct.
Answer B is wrong. Technician A is also correct.
Answer C is correct. Both technicians are correct. A digital diagnostic reader (DDR) can both read active codes and erase historic (inactive) codes.
Answer D is wrong. Both technicians are correct.

Question #12
Answer A is wrong. Replacing the breather is not cleaning the breather.
Answer B is wrong. Soaking the breather in gasoline is not a recommended procedure.
Answer C is correct. The best way to clean a rear-axle breather is to soak in solvent and then use compressed air.
Answer D is wrong. Wiping the orifice will not clean any dirt trapped inside the breather.

Question #13
Answer A is wrong. The temperature of the engine is not important for this test.
Answer B is wrong. The temperature of the engine is not important for this test.
Answer C is correct. When checking for fuel leaks on an engine, it is mandatory that the engine be operating so the fuel system is pressurized, thus making a leak very apparent.
Answer D is wrong. The engine must be running to pressurize the fuel system for the test.

Question #14
Answer A is wrong. Technician B is also correct.
Answer B is wrong. Technician A is also correct.
Answer C is correct. Both technicians are correct. You connect the battery positive cable first when hooking up a battery. One always disconnects the ground cable first when disconnecting a battery.
Answer D is wrong. Both technicians are correct.

Question #15
Answer A is wrong. PMI records on heavy trucks are required by the DOT.
Answer B is correct. Only Technician B is correct. Maintenance records should be referenced before undertaking major repairs.
Answer C is wrong. Only Technician B is correct.
Answer D is wrong. Only Technician B is correct.

Question #16
Answer A is correct. Starter draw testing is not part of a road test procedure.
Answer B is wrong. Brake air pressure and leakage is part of road test procedures.
Answer C is wrong. Tire checking is part of road test procedures.
Answer D is wrong. Checking for unusual noises is also a part of road testing.

Question #17

Answer A is wrong. Technician B is also correct.

Answer B is wrong. Technician A is also correct.

Answer C is correct. Both technicians are correct. You can use an inductive clamp ammeter to check alternator output. It is normal for alternator output to slightly decrease as ambient temperature increases.

Answer D is wrong. Both technicians are correct.

Question #18

Answer A is correct. Only Technician A is correct. All current (manufactured after 1998) tractors and trailers are required to have automatic slack adjusters.

Answer B is wrong. Automatic slack adjusters do not increase the allowable free stroke dimension.

Answer C is wrong. Only Technician A is correct.

Answer D is wrong. Only Technician A is correct.

Question #19

Answer A is wrong. Although tires are a high wear item, they account for 17.6 percent of complaints.

Answer B is wrong. Lights account for 14.5 percent of all maintenance concerns.

Answer C is wrong. Suspension concerns account for only 6.2 percent of maintenance concerns.

Answer D is correct. Brakes are the highest cited area on a PMI. Brakes account for 25.9 percent of maintenance concerns and top the list of maintenance concerns that fleets cite when asked to name the top five trailer maintenance problems.

Question #20

Answer A is wrong. A leaking air spring will cause the trailer to sit low, not high.

Answer B is correct. Only Technician B is correct. An improperly adjusted height control valve could be the problem.

Answer C is wrong. Only Technician B is correct.

Answer D is wrong. Only Technician B is correct.

Question #21

Answer A is wrong. A seized floating caliper on a hydraulic brake system does not cause brake pedal fade.

Answer B is wrong. A seized floating caliper on a hydraulic brake system does not cause brake pedal pulsations.

Answer C is correct. A seized floating caliper on a hydraulic brake system causes reduced braking force.

Answer D is wrong. A seized floating caliper on a hydraulic brake system does not cause brake grabbing.

Question #22

Answer A is correct. Loose mounting bolts would cause the chamber to move when the brakes are applied, but would be unlikely to cause an audible leak.

Answer B is wrong. A ruptured diaphragm could cause an air leak.

Answer C is wrong. A loose chamber clamp could cause an air leak.

Answer D is wrong. A missing clamp bold could cause an air leak.

Question #23

Answer A is wrong. The vehicle should be on level ground at a distance of 25 feet, not 35 feet as stated.

Answer B is wrong. You do not always use headlight aiming equipment when adjusting headlights. One can use a screen or chart.

Answer C is wrong. Neither technician is correct.

Answer D is correct. Neither technician is correct.

Question #24

Answer A is correct. Leaking, worn-out front shock absorbers can cause the steering wheel to shake for a few seconds after a stretch of rough road.
Answer B is wrong. Low tire pressure would not cause wheel shake.
Answer C is wrong. Rusted rear shock absorbers would not cause wheel shake.
Answer D is wrong. A missing jounce bumper would cause a bang on a rough road but not a shake.

Question #25

Answer A is wrong. A short to ground through the starter motor would cause a high current draw.
Answer B is wrong. Any mechanical resistance would require more energy, thus more current.
Answer C is correct. Small gauge starter wire will cause high resistance and low current draw.
Answer D is wrong. A short to ground in the starter wire would cause increased current flow.

Question #26

Answer A is wrong. 45 degrees upward from level would increase the inflation of the air springs.
Answer B is correct. When the valve is in the neutral position, the lever should be level.
Answer C is wrong. 45 degrees downward from level would decrease the inflation of the air springs.
Answer D is wrong. This position is not possible.

Question #27

Answer A is wrong. Foreign particles on the valve seat could cause a constant air leak from the dryer purge valve.
Answer B is wrong. A damaged purge valve seat could cause a constant air leak from the dryer purge valve.
Answer C is wrong. A frozen purge valve could cause a constant air leak from the dryer purge valve.
Answer D is correct. A restricted line will NOT cause a constant air leak from the dryer purge valve; therefore, it is the exception.

Question #28

Answer A is wrong. A rotten egg smell in the cab is caused by a plugged evaporator case drain.
Answer B is wrong. A rotten egg smell in the cab is caused by a plugged evaporator case drain.
Answer C is wrong. A rotten egg smell in the cab is caused by a plugged evaporator case drain.
Answer D is correct. A rotten egg smell in the cab is caused by a plugged evaporator case drain.

Question #29

Answer A is correct. Vinegar will kill the mold and mildew that are the source of the odor.
Answer B is wrong. This method will simply mask the odor and not eliminate the problem.
Answer C is wrong. This method poses a possible fire hazard.
Answer D is wrong. This method will simply mask the odor and not eliminate the problem.

Question #30

Answer A is wrong. An exhaust leak under the cab or sleeper is a CVSA out-of-service condition.
Answer B is wrong. Incorrect routing of a wiring harness, air lines, or fuel lines allowing them to contact the exhaust are CVSA out-of-service conditions.
Answer C is correct. An inaccurate exhaust temperature gauge would not by itself place a truck out of service.
Answer D is wrong. Missing heat shields and guards could deadline a truck.

Question #31

Answer A is wrong. Technician B is also correct.
Answer B is wrong. Technician A is also correct.
Answer C is correct. Both technicians are correct. Semi-elliptical springs are usually mounted with a pin and bushing to the spring hanger and the other end is free to move forward and backward to compensate for changes in length. Multiple leaf springs should be replaced as an assembly.
Answer D is wrong. Both technicians are correct.

Question #32
Answer A is wrong. Technician B is also correct.
Answer B is wrong. Technician A is also correct.
Answer C is correct. Both technicians are correct. Checking the operation of the clutch and gearshift is part of a drive test during a PMI. When a drive test is performed, the speed at which the governor is set is always checked.
Answer D is wrong. Both technicians are correct.

Question #33
Answer A is wrong. 39 inches is not considered a standard fifth-wheel height.
Answer B is wrong. 43 inches is not considered a standard fifth-wheel height.
Answer C is correct. 47 inches is considered the standard fifth-wheel height. Most trailers are designed with an intended 47 inches upper-coupler (kingpin) plate height. This, plus the fact that any trailers operated on North American roads cannot exceed a height of 13 feet, 6 inches, determines the plate height of fifth wheels.
Answer D is wrong. 49 inches is not considered a standard fifth-wheel height.

Question #34
Answer A is wrong. Constant rate springs are leaf-type assemblies that have a constant rate of deflection.
Answer B is wrong. Progressive rate springs are leaf-type spring assemblies with a variable deflection rate obtained by varying the effective length of the spring assembly.
Answer C is wrong. Auxiliary springs are leaf-type spring assemblies usually mounted on top of the truck rear spring assemblies to increase load-carrying ability.
Answer D is correct. Coil springs are the exception because they are NOT used on heavy-duty truck suspensions.

Question #35
Answer A is wrong. An over-pressed pulley could cause the power-steering-pump pulley to become misaligned.
Answer B is wrong. A loose fit from the pulley hub to pump shaft could cause the power-steering-pump pulley to become misaligned.
Answer C is wrong. A worn or loose pump mounting bracket could cause the power-steering-pump pulley to become misaligned.
Answer D is correct. A broken engine mount will NOT cause the power-steering-pump pulley to become misaligned.

Question #36
Answer A is wrong. Performing a continuity test (ohms) involves using the power source of the meter to determine resistance. Source voltage should be disconnected before using an ohmmeter.
Answer B is correct. Only Technician B is correct. Using voltage drop is a an accurate means of identifying high resistance in an energized circuit.
Answer C is wrong. Only Technician B is correct.
Answer D is wrong. Only Technician B is correct.

Question #37
Answer A is wrong. A cooling system leak is not considered a safety priority.
Answer B is correct. Only Technician B is correct. There must be sufficient vacuum for at least one full brake application after the engine is shut off, otherwise it must be deadlined.
Answer C is wrong. Only Technician B is correct.
Answer D is wrong. Only Technician B is correct.

Question #38
Answer A is wrong. Technician B is also correct.
Answer B is wrong. Technician A is also correct.
Answer C is correct. Both technicians are correct. S-cam-actuated foundation brake systems are the most common type of foundation brake system found on heavy-duty trucks. Slack adjusters working with S-cams convert the linear force developed by the brake chamber into brake torque.
Answer D is wrong. Both technicians are correct.

Question #39
Answer A is wrong. A loose feeling in steering wheel when driving straight can be caused by a worn front steering bushing.
Answer B is wrong. A noise when hitting a bump in the road can be caused by a worn front steering bushing.
Answer C is correct. Excessive use of power steering fluid is the exception and cannot be caused by a worn steering bushing.
Answer D is wrong. A pull to the left while driving on the road can be caused by a worn front steering bushing.

Question #40
Answer A is wrong. Technician B is also correct.
Answer B is wrong. Technician A is also correct.
Answer C is correct. Both technicians are correct. The vibration damper counterbalances the back-and-forth twisting motion of the crankshaft each time a cylinder fires. If the seal contact area on the vibration damper hub is scored, the damper assembly must be replaced.
Answer D is wrong. Both technicians are correct.

Question #41
Answer A is wrong. A bell-mouth condition results when the drum diameter is greater at the edge of the drum next to the backing plate compared to the edge near the wheel.
Answer B is wrong. A concave brake drum results when the drum diameter is greater in the center compared to both edges.
Answer C is wrong. An out-of-round condition results when there is a variation in readings 180 degrees apart.
Answer D is correct. A convex drum results when the diameter on a brake drum is greater at the edges of the friction surface than in the center.

Question #42
Answer A is wrong. The cooling system does not use this type of valve.
Answer B is correct. The exhaust system uses a sliding gate valve in the exhaust brake. The sliding gate exhaust brake uses a pneumatically activated gate actuated by chassis system pressure. The air supply to close the gate is controlled by an electronically switched pilot valve. An aperture in the sliding gate permits minimal flow through the brake gate during engine braking. The butterfly valve version operates similarly.
Answer C is wrong. The A/C system does not use this type of valve.
Answer D is wrong. The electrical system does not use this type of valve.

Question #43
Answer A is wrong. A one-way clutch is not used in a driveline.
Answer B is wrong. There is no such part as a vertical yoke support.
Answer C is wrong. The exhaust pipe supports do not have that appearance.
Answer D is correct. A center support bearing is shown. Center support bearings are used when the distance between the transmission and the rear axle is too great to span with a single driveshaft. The center bearing is fastened to the frame and aligns the two connecting driveshafts. It consists of a stamped steel bracket used to align and fasten the bearing to the frame. A rubber mount inside the bracket surrounds the bearing, which supports the driveshaft.

Question #44
Answer A is wrong. Technician B is also correct.
Answer B is wrong. Technician A is also correct.
Answer C is correct. Both technicians are correct. The frame rail marked "A" is an incorrect repair since the ends should be cut at a 45 or 60 degree angle and extend beyond the stress point. The frame reinforcement on the frame rail marked "B" is called an external channel reinforcement.
Answer D is wrong. Both technicians are correct.

Question #45
Answer A is correct. Only Technician A is correct. The pedal free travel should be about 1½ to 2 inches (38.1 to 50.8 mm). This clearance ensures full clutch engagement, and no interference from the linkage should occur.
Answer B is wrong. If the release bearing to clutch brake travel is less than ½ inch the release bearing will contact the clutch brake too soon, causing clutch brake damage, and complete clutch disengagement may not be possible.
Answer C is wrong. Only Technician A is correct.
Answer D is wrong. Only Technician A is correct.

Question #46
Answer A is wrong. Technician B is also correct.
Answer B is wrong. Technician A is also correct.
Answer C is correct. Both technicians are correct. An engine should be checked while doing an inspection for both static and dynamic leaks. This means that the engine should be checked while it is running and while it is off. This ensures that all gasket and mating surfaces are monitored in all possible operating and nonoperating conditions. When checking for fuel leaks on an engine, it is mandatory that the engine be operating so the fuel system is pressurized, thus making a leak very apparent.
Answer D is wrong. Both technicians are correct.

Question #47
Answer A is wrong. Informing the driver on such an important safety issue is not enough.
Answer B is wrong. An important safety problem such as brakes requires the vehicle to be deadlined until proper repairs are made and the work is noted using a sign-off sheet before the vehicle is put back into service. CVSA out-of-service criteria shows that brake linings that are saturated with oil, grease, or brake fluid is an out-of-service condition.
Answer C is wrong. Neither technician is correct.
Answer D is correct. Neither technician is correct.

Question #48
Answer A is wrong. It is normal (heating and cooling) for drive axles to acquire slight condensation (which usually "boils off" during normal driving conditions).
Answer B is correct. Infrequent driving and short trips do not bring the axle to temperature and condensed moisture does not boil off.
Answer C is wrong. Submerging the axles in water can introduce water into the axle, but it is not the most likely cause.
Answer D is wrong. Any water that happens to make its way into the axle would normally be "boiled off" during normal driving conditions.

Question #49
Answer A is wrong. Leakage at the brake application valve exhaust port with the brakes applied should not exceed a 1-inch (2.54 mm) bubble in 3 seconds.
Answer B is wrong. Leakage at the brake application valve exhaust port with the brakes applied should not exceed a 1-inch (2.54 mm) bubble in 3 seconds.
Answer C is wrong. Leakage at the brake application valve exhaust port with the brakes applied should not exceed a 1-inch (2.54 mm) bubble in 3 seconds.
Answer D is correct. Leakage at the brake application valve exhaust port with the brakes applied should not exceed a 1-inch (2.54 mm) bubble in 3 seconds.

Question #50

Answer A is correct. When the ABS check switch and the ignition switch are turned on, the ABS light flashes blink codes.

Answer B is wrong. The first set of flashes indicates the ABS configuration.

Answer C is wrong. When there are no electrical defects in the ABS, a 1,00, 2,00, or 4,00 blink code is provided.

Answer D is wrong. A blink code indicates a defect in a certain area, but not necessarily in a specific component.

Question #51

Answer A is wrong. Only DOT 3 brake fluid should be used in a hydraulic clutch master cylinder.

Answer B is correct. Hydraulic clutch master cylinders use DOT 3 brake fluid. The rubber seals in this system are incompatible with the other fluids suggested. Brake fluid is used because it provides a controlled amount of seal swell to supply proper system sealing, can withstand extreme temperature extremes, is compatible with rubber seals, resists evaporation even at high temperatures, and combats the formation of rust and corrosion in the system.

Answer C is wrong. Only DOT 3 brake fluid should be used in a hydraulic clutch master cylinder.

Answer D is wrong. Only DOT 3 brake fluid should be used in a hydraulic clutch master cylinder.

Question #52

Answer A is wrong. The rear U-joint is not lubricated from that location.

Answer B is correct. Lubricating the slip joint is what is being shown in the figure. To lubricate the slip joint, apply grease gun pressure to the lube fitting until lubricant appears at the pressure relief hole in the plug at the slip yoke end of the spline. Now cover the pressure relief hole with your finger and continue to apply pressure until grease appears at the slip yoke seal. Sometimes it might be easier to purge the slip yoke by removing the dust cap and reinstalling it once grease appears.

Answer C is wrong. The center bearing is not lubricated from that location.

Answer D is wrong. The rear bearing (if applicable) is not lubricated from that location.

Question #53

Answer A is wrong. A stud terminal maintenance-free battery is shown.

Answer B is correct. A stud terminal maintenance-free battery is shown. Top stud terminals incorporate a threaded terminal design. Side terminal batteries use internal threaded posts positioned in the side of the battery case that require a special bolt to connect the cables. Post terminal batteries use round, tapered terminals on top of the battery. The positive post is larger in diameter than the negative post. This helps to identify the polarity of the battery post and prevent connecting the battery in reverse polarity.

Answer C is wrong. A stud terminal maintenance-free battery is shown.

Answer D is wrong. A stud terminal maintenance-free battery is shown.

Question #54

Answer A is wrong. Replacing the lost coolant with water or antifreeze would result in an incorrect reading and does not follow the proper sequence of servicing the cooling system.

Answer B is wrong. Cleaning the coolant is performed after testing the coolant's condition.

Answer C is wrong. Adding antifreeze to the system should only be performed after a full evaluation of the existing coolant has been performed.

Answer D is correct. Test the condition of the coolant first. Coolant freeze and boil protection levels can be tested with a hydrometer or a refractometer. Different antifreeze and water ratios are required to raise or lower the protection level depending upon the temperatures the vehicle will encounter. A 50/50 mixture is most common.

Question #55

Answer A is wrong. Mechanical advantage is a design factor and not routinely checked by technicians.
Answer B is wrong. Clutch linkage angles are design factors and are not routinely checked by technicians.
Answer C is wrong. Clutch linkage clevis leverage is a design factor and not routinely checked by technicians.
Answer D is correct. Checking clutch linkage while performing preventative maintenance would include checking for binding or wear. The clutch will not operate correctly if the linkage is worn or damaged. Make sure the linkage is not obstructed. Make sure every pivot point in the linkage operates freely. Make sure the pedal, springs, brackets, bushings, shafts, clevis pins, levers, cables, and rods are not worn or damaged. Do not straighten any damaged parts; replace the part. Make sure every pivot point in the linkage is lubricated.

Question #56

Answer A is correct. The condition of fuel tanks, mountings, lines, and caps are a concern during a PMI. The fuel tank mounts should be securely attached to the frame and the tank and be free from any cracks, corrosion, and damage that could cause future failure. There should be no signs of tank shifting and all mounting bolts should be in place and tight. Tanks must be checked for any signs of leakage at any point, with special attention to all seams and any damaged areas. Check to ensure that the cap seal is in place and in good condition.
Answer B is wrong. Pour point tests are never performed as part of a PMI.
Answer C is wrong. The heater lines have nothing to do with a fuel system PMI.
Answer D is wrong. Fuel gauges are not recalibrated as part of PMI.

Question #57

Answer A is correct. Only Technician A is correct. A turbocharger should be checked for oil leaks when the engine is running. Without the engine producing oil pressure, it would be more difficult to detect leaks. The oil pressure to a turbocharger feeds the bearings, spills to an oil drain cavity, and then drains to the crankcase by means of a return hose.
Answer B is wrong. It would be dangerous and impossible to try to check the turbocharger shaft for runout while the engine is running. When running a turbocharger-equipped diesel engine without the air intake, be sure to install an intake guard over the turbocharger inlet to avoid personal injury from the exposed rotating impeller.
Answer C is wrong. Only Technician A is correct.
Answer D is wrong. Only Technician A is correct.

Question #58

Answer A is wrong. Technician A immediately greases the hinges although that may not be the cause of the door not closing properly. Many factors could contribute to the door not closing properly.
Answer B is correct. Only Technician B is correct. Listening to or reading the driver's reports and then checking the truck's maintenance records for past problems or repairs before proceeding is the proper diagnostic sequence to follow. A thorough inspection of the door and associated components such as hinges and latches should be performed before determining a proper course of action.
Answer C is wrong. Only Technician B is correct.
Answer D is wrong. Only Technician B is correct.

Question #59

Answer A is wrong. Technician B is also correct.
Answer B is wrong. Technician A is also correct.
Answer C is correct. Both technicians are correct. Underground fuel storage tanks can form condensation and should always be checked for water content. Checks need to be diligent and any water found in the fuel system must be removed. A battery that is even partially discharged will be exposed to a freeze-up problem in cold weather. The specific gravity in a discharged battery is decreased to a point where the electrolyte can freeze. Electrolyte in a fully charged battery has a specific gravity of 1.265 and will not freeze until it is approximately below 0°F. A discharged battery with an electrolyte specific gravity of 1.1 will freeze at approximately 15°F.
Answer D is wrong. Both technicians are correct.

Question #60
Answer A is wrong. Technician B is also correct.
Answer B is wrong. Technician A is also correct.
Answer C is correct. Both technicians are correct. The vehicle should be deadlined if weld repairs are found to be made on pintle chains and hooks. CVSA vehicle out-of-service criteria includes cracks anywhere in the pintle hook assembly or any visible welded repairs to the pintle hook. Out-of-service criteria for pintle chains and hooks also lists improper repairs to chains and hooks including welding, wire, small bolts, rope, and tape. Loads projecting more than 4 feet beyond the vehicle body must have at least one operative red or amber light on the rear of loads visible from 500 feet.
Answer D is wrong. Both technicians are correct.

Question #61
Answer A is wrong. Ballast resistors are used for ignition systems.
Answer B is correct. Only Technician B is correct. You should check the cab ground integrity. Interior cab lights and circuits are affected by resistance problems such as loose terminals, corroded connectors, and high-resistance ground connections. Proper grounding of the dash circuits and interior lighting should always be checked whenever a problem is suspected. When more than one light has similar symptoms, like not operating or being dim, the cab may not be adequately grounded. This is especially true when lights are attached to plastic panels, and the ground must be completed elsewhere.
Answer C is wrong. Only Technician B is correct.
Answer D is wrong. Only Technician B is correct.

Question #62
Answer A is wrong. The speedometer monitors and indicates road speed.
Answer B is wrong. A fuel pressure gauge monitors and indicates fuel pressure.
Answer C is correct. The pyrometer measures exhaust gas temperature. The pyrometer is normally found at a specified distance down line from the turbocharger and can provide the driver with engine loading information useful for shifting. Thermocouple pyrometers are used. This consists of two dissimilar insulated wires (often pure iron and constantin [55 percent copper and 45 percent nickel]) connected at each end to form a continuous circuit. The two junctions are known as the hot end, located where temperature reading is required, and the reference end, which is connected to a millivolt meter. Whenever the junctions are at different temperatures, there will be current flow; the voltage will increase with a greater difference in temperature.
Answer D is wrong. The ammeter monitors and indicates amperes.

Question #63
Answer A is wrong. Cold-temperature operation can cause transmission and axle lubrication problems know as channeling. High-viscosity oils typically used in transmissions and axles thicken and may not flow at all when the vehicle is first started. Rotating gears can push lubricant aside, leaving voids or channels where no lubricant is actually touching the gears.
Answer B is correct. Over-greasing can be as damaging as under-greasing and can cause blowing out of a bearing seal. Grease guns exert tremendous pressure and oozing may mean that a bearing seal, if used, has blown out. When this happens, the lubricant will be contaminated and lead to failure of the part.
Answer C is wrong. Scoring is caused by insufficient lubrication due to insufficient volume, which can lead to part or component failure.
Answer D is wrong. An accumulation of grit around greasing points is a normal occurrence. Wipe off excess grease to avoid attracting dust and grit in and around the lubrication point.

Question #64
Answer A is wrong. Technician B is also correct.
Answer B is wrong. Technician A is also correct.
Answer C is correct. Both technicians are correct. A carbon pile load tester can be used to perform an alternator output voltage and amperage test. Connect it across the battery, observing correct polarity. Start the engine and run at about 1500 rpms. Observe the unloaded output voltage. It should be between 13.5 to 14.5 volts. Attach the inductive pick-up to the alternator output wire. With the engine at high idle, use the load tester to draw the highest amperage reading without dropping the voltage to fall below 12 volts. This procedure forces the alternator to increase its output to near its maximum. The ammeter should indicate output within 5 percent of the alternator's capacity.
Answer D is wrong. Both technicians are correct.

Question #65
Answer A is wrong. Technician B is also correct.
Answer B is wrong. Technician A is also correct.
Answer C is correct. Both technicians are correct. A grounded sending unit wire may cause the gauge to fail from excess current flow. A short circuit in the sending unit wire to the gauge will lower resistance in that circuit. Generally a short to ground in a sending unit wire will cause an overload of current to the gauge, which will consequently burn out in a very short time. Trucks with electronic gauges perform a self-test by sweeping the gauges when the ignition switch is turned on. A proper and thorough check of a vehicle's instrument panel is always an important part of any PMI. Most vehicles also have a self-check feature that illuminates all warning lamps when the ignition switch is first switched to the on position.
Answer D is wrong. Both technicians are correct.

Question #66
Answer A is wrong. On the PMI you do have to check the cab-raising hydraulic cylinders for leaks. Inspect the hydraulic cylinders for external leakage and determine if the cab can be raised and lowered. Check for internal leakage by observing bleed down. Make sure the mechanical stops engage and hold safely.
Answer B is correct. Only Technician B is correct. On the PMI you must check to see if the cab will raise and lower properly. The push circuit raises the cab from the lowered position to the desired tilt position. The pull circuit brings the cab from a fully tilted position up and over the center. Remember that in most systems, whenever raising or lowering the cab, stop working the hydraulic pump once the cab goes over center. The cab falls at a controlled rate and continued pumping could lock up the tilt cylinder.
Answer C is wrong. Only Technician B is correct.
Answer D is wrong. Only Technician B is correct.

Question #67
Answer A is correct. A visual inspection will not reveal the proper belt tension. A belt tension gauge should be used whenever possible for the most accurate readings. If one is not available, then the normal deflection points at the specified locations and measurements should be observed.
Answer B is wrong. Premature wear due to misalignment can be seen on a visual inspection. Belt pulleys must be properly aligned to minimize belt wear. The edges of the pulleys must be in line when a straightedge is placed on the pulleys.
Answer C is wrong. Correct orientation of a dual belt application can be seen on a visual inspection.
Answer D is wrong. Proper belt seating in the pulley groove can be seen on a visual inspection. Because the friction surfaces are on the side of a V-belt, wear occurs in this area. If the belt edges are worn, the belt may be rubbing on the bottom of the pulley. The lower edge should never touch the bottom of the pulley groove.

Question #68

Answer A is wrong. The Federal Highway Administration has set up a minimum inspection standards program, under which each vehicle must carry proof that an inspection was completed. Proof can either be a copy of the inspection form kept on the vehicle or a decal. If using a decal, a copy of the inspection form must be kept on file and the decal must indicate where an inspector can get a copy of it.

Answer B is wrong. Each vehicle must be inspected separately. This means a tractor/trailer and/or converter dolly is considered to be three vehicles, each requiring a decal or inspection form. Decals must be legible and show the following information: date vehicle passed inspection, name and address to contact about inspection records, and VIN.

Answer C is wrong. Neither technician is correct.

Answer D is correct. Neither technician is correct.

Question #69

Answer A is wrong. Technician B is also correct.

Answer B is wrong. Technician A is also correct.

Answer C is correct. Both technicians are correct. If the alternator mounting holes are distorted due to overtorquing and/or age, then the alternator should be replaced. Severely out-of-round mounting holes on an alternator will cause misalignment of the drive belt and may cause premature bearing failure. A complaint of an unusual noise requires a thorough visual inspection. A visual inspection will usually determine whether a loose component is causing a vibration or any unusual sounds. Before performing a system output test, the technician must first check for a loose alternator belt that could cause a pulley to slip, resulting in low alternator output. If a belt is worn or glazed, it will prevent the alternator from attaining the proper rpm due to slippage. A belt tension gauge should be used whenever possible for the most accurate readings.

Answer D is wrong. Both technicians are correct.

Question #70

Answer A is wrong. The operation of a road speed governor cannot be checked on a pre-trip inspection. Checking the road speed limiter is part of a vehicle road test. Connect the appropriate engine scan tool, OEM special tool, or laptop computer to the truck's diagnostic connector. During the road test, activate the tool's parameter specification readout and record the road speed. Reprogram if required.

Answer B is wrong. Although the regulations seem to place greater emphasis on the post-trip inspection, most operators agree the pre-trip inspection is most important. Because the driver doing the inspection will be operating that particular vehicle, the incentive to ensure that the vehicle is safe is probably greater. Repairs are cheaper and less time-consuming if the driver finds them on the pre-trip before inspectors levy fines or impose penalties.

Answer C is wrong. Neither technician is correct.

Answer D is correct. Neither technician is correct.

Question #71

Answer A is wrong. The pour point of a fuel is generally considered to be 5°F above the temperature at which the fuel will no longer flow. As the temperature is lowered below the cloud point, more and more of the paraffinic compounds of the fuel crystallize. As they do, they begin to combine to form a solid gel-like structure, which cannot be poured or pumped. Pour point depressant additives are required for extreme cold weather operation in #1D and #2D diesel fuel. Pour point depressants have no effect on cloud point.

Answer B is correct. The cloud point is the temperature at which the normal paraffin in a fuel becomes less soluble and begins to precipitate as wax crystals and make the fuel appear cloudy and hazy. The cloud point exceeds the pour point by 5°F (3°C) to 25°F (15°C).

Answer C is wrong. The point at which evaporation begins is not a factor to consider in normal and/or severe operating conditions.

Answer D is wrong. Black smoke is produced when excess fuel is burned.

Question #72
Answer A is wrong. Two people are necessary to perform a proper steering system check.
Answer B is wrong. The belts are adjusted before any tests are performed.
Answer C is wrong. Neither technician is correct.
Answer D is correct. Neither technician is correct. The best method for checking a steering system is with the wheels on the floor and the engine running. Allow the engine to idle with the transmission in neutral and the parking brakes applied. While someone turns the steering wheel one-quarter turn in each direction from the straight-ahead position, observe all the pivoted joints on the tie-rod ends and drag link and other steering systems-related components. This allows the technician to check the steering linkage pivots under load. Manual steering wheel play should not exceed 2 inches and power steering wheel play should not exceed 2½ inches. All belts should be checked for obvious signs of wear, tension, and/or correct application. A belt tension gauge should be used whenever possible for the most accurate readings.

Question #73
Answer A is correct. Only Technician A is correct. Whenever any repair procedure or adjustment is performed on a vehicle, the work must be noted using a sign-off sheet before the vehicle is put back into service. If the problem noted did not require repair, but was still looked at, then this must be recorded as well. A copy of the inspection report must remain with the vehicle.
Answer B is wrong. A PMI worksheet is not the correct place to record repairs. PMI worksheets are used for work that is scheduled in advance. The following is a list of basic types of PMI inspections: Schedule "A" is a light inspection. Schedule "B" is a more detailed inspection. Schedule "C" is a detailed inspection, service, and adjustment. Schedule "D" is a comprehensive inspection and adjustment.
Answer C is wrong. Only Technician A is correct.
Answer D is wrong. Only Technician A is correct.

Question #74
Answer A is correct. Lateral runout and rotor thickness measurements are shown in the figure. To check lateral runout, mount a dial indicator on the steering arm or anchor plate with the indicator plunger contacting the rotor 1 inch from the edge of the rotor. Check lateral runout on both sides of the rotor. Zero the indicator and measure the total indicated runout (TIR). The lateral runout TIR is always a tight dimension, seldom exceeding 0.015 inch. If lateral runout exceeds the specifications, the rotor will have to be either machined or replaced. Thickness variation should also be measured at 12 equidistant points with a micrometer at about 1 inch from the edge of the rotor. If thickness measurements vary by more than the specified maximum, usually around 0.002 inch, the rotor should be machined or replaced.
Answer B is wrong. Lateral runout and rotor thickness measurements are shown in the figure.
Answer C is wrong. Lateral runout and rotor thickness measurements are shown in the figure.
Answer D is wrong. Lateral runout and rotor thickness measurements are shown in the figure.

Question #75
Answer A is wrong. Technician B is also correct.
Answer B is wrong. Technician A is also correct.
Answer C is correct. Both technicians are correct. A crack in a converter dolly safety chain is a CVSA out-of-service (OOS) condition requiring deadlining of the vehicle. Rear chains and hooks that are used in double trailer systems for hookup should not be worn to the extent of a measurable reduction in link cross-sectional area or have any significant abrasions, cracks, or other faults that would affect structural integrity. Improper repairs to chains and hooks including welding, wire, small bolts, rope, and tape are also OOS conditions. A worn drawbar eye is also an OOS condition. No part of the eye should have any section reduced by more than 20 percent and no section reduction should be visible when coupled. OOS criteria also include fasteners missing or ineffective, mounting cracks in attachment welds, and any visible cracks in the drawbar.
Answer D is wrong. Both technicians are correct.

Question #76

Answer A is wrong. Coolant is not color-coded by type.

Answer B is wrong. You only check the cooling system at full capacity.

Answer C is wrong. Neither technician is correct.

Answer D is correct. Neither technician is correct. It is important to know that all coolants are not alike and are not always compatible with each other. Identify the coolant being used and be sure to follow the pertinent change intervals and refill with the correct type. ELCs have a longer service life than EG and PG coolants. It is important to test the radiator cap to ensure that it will retain the indicated pressure setting. Using a pressure tester unit, install the cap onto the tester and apply pressure to the specified pressure release point. If the specified limit cannot be achieved, then cap replacement is required. Pressure testing the engine cooling system for leaks is performed in the same manner. Apply pressure by means of the same tester unit. The gauge reading should achieve the same setting as the recommended radiator cap. The gauge reading should hold steady.

Question #77

Answer A is wrong. A few particles indicate normal wear, not a problem needing immediate resolution. An experienced technician can usually determine the operating condition of the drive axle differential by the fluid. Most drive axles are equipped with a magnetic drain plug that is designed to attract any metal particles suspended in the gear oil. A nominal amount of "glitter" is normal because of the high torque environment of the drive axle. However, too much "glitter" indicates a problem that requires further investigation.

Answer B is correct. The technician needs to inform the customer of the condition and tell them to monitor the amount of particles.

Answer C is wrong. It is always best to inform the customers of any condition on their vehicle that may require extra attention.

Answer D is wrong. A few particles indicate normal wear, not a problem needing immediate resolution.

Question #78

Answer A is wrong. A driveline vibration problem could originate from a new or used driveshaft. Excessive binding on the U-joint crosses will cause a disruption in the overall harmonics of the system, making vibration evident. A change in U-joint operating and/or canceling angles will affect shaft vibration. The transmission output shaft and rear-axle input shaft should be checked because they can affect the shaft's harmonics. End yokes should be checked for retaining nut looseness, radial looseness, fractures, runout, and vertical alignment, which could cause vibration. Inspect for slip spline slop, center bearing problems, and improper phasing.

Answer B is wrong. The problem is unlikely to originate in the transmission considering the high vibration is only encountered at high speeds. Vibrations at high speeds are often indicative of a driveline problem.

Answer C is wrong. Neither technician is correct.

Answer D is correct. Neither technician is correct.

Question #79

Answer A is wrong. Technician B is also correct.

Answer B is wrong. Technician A is also correct.

Answer C is correct. Both technicians are correct. A battery that has recently been charged by running the engine, or using a battery charger, must have its surface charge removed before performing a battery load test. The surface charge may be removed by applying a load of 300 amps for 15 seconds with a carbon pile load tester, or by cranking the engine for 15 seconds. After removing the surface charge, stabilize the voltage by allowing the battery to rest for 15 minutes. If multiple batteries connected, then first disconnect the batteries and perform individual testing on each battery. If the voltage is below 12.4 volts DC at 70°F, the battery should be recharged to above 12.6 volts. Apply a load through the tester at one-half the cold-cranking rating (CCA) for 15 seconds. If the voltage drops below 9.6 volts at 70°F, the battery is unserviceable.

Answer D is wrong. Both technicians are correct.

Question #80
Answer A is wrong. Technician B is also correct.
Answer B is wrong. Technician A is also correct.
Answer C is correct. Both technicians are correct. The type of fire extinguisher carried on a truck should be determined by factors such as the type of cargo carried. Most fire extinguishers are marked by a letter and symbol to indicate the classes of fires for which they can be used. Every truck or truck-tractor with a gross vehicle weight rating (GVWR) of 10,001 pounds or more must have a fire extinguisher. If the vehicle is used for transporting hazardous material that requires placards, the extinguisher must have an Underwriter's Laboratory (UL) rating of 10B:C or more. If the vehicle is not used for hazardous materials, an extinguisher with a UL rating of 5B:C may be used. The rating is usually shown near the UL certification on the extinguisher. Section 396.19 of the Federal Motor Carrier Safety Regulations states it is the fleet manager's responsibility to ensure that all personnel doing annual inspections are qualified to perform those inspections. They should understand the inspection criteria and be able to identify defective components and systems.
Answer D is wrong. Both technicians are correct.

Question #81
Answer A is wrong. Any damage or dents to the driveshaft tube will cause a change in its harmonics and lead to vibration problems. Check the shaft for damaged, bent tubing or missing balance weights. Make certain there is no buildup of foreign material on the shaft, such as undercoat or concrete.
Answer B is wrong. The universal joint allows the driveshaft to operate at different and constantly changing angles, and excessive binding on the crosses will cause a disruption in the overall harmonics of the system, making vibration evident.
Answer C is wrong. Changes in operating and/or canceling angles will affect shaft vibration. Any given universal joint has a maximum angle at which it will still transmit torque smoothly. This angle depends in part on the joint size and design. Exceeding the maximum recommended working angle will greatly shorten or immediately destroy the joint service life. Ideally, the operating angles on each end of the drive shaft should be equal to or within 1 degree of each other, have a 3-degree maximum operating angle, and have at least ½ degree of continuous operating angle. Having the transmission output shaft and axle input shaft at relatively similar angles, thus resulting in equal U-joint angles, is known as driveshaft U-joint cancellation.
Answer D is correct. Rust on the driveshaft tube is inconsequential and will not cause driveshaft vibration.

Question #82
Answer A is wrong. Technician B is also correct.
Answer B is wrong. Technician A is also correct.
Answer C is correct. Both technicians are correct. When considering whether to deadline a truck, the importance of moving a load must be weighed against the probability that a mechanical condition will result in a breakdown. The Federal Motor Carrier Safety Regulations, Part 393, say that anything mechanical that can either cause or prevent an accident is a safety item. All alleged safety problems should be addressed. The regulations also forbid dispatching a vehicle that is likely to break down. Regardless of whether any fine might be levied for that, dispatching a truck that is on the brink of breaking down is just not good business; nor is it safe. Items such as a malfunctioning cruise control or an engine brake are not deadline items and would not warrant deadlining the truck. A proper and thorough check of a vehicle's instrument panel is always an important part of any PMI. Most vehicles have a self-check feature that illuminates all warning lamps when the ignition switch is first switched to the on position. Warning lights and buzzers for oil, coolant, and charging circuit should go out right away. The low air pressure alarm (which is mandatory) must sound when the system pressure drops below 50 percent of the governor cut-out pressure (approximately 60 psi). When the ignition is first turned on, electronic gauges perform a self-test by sweeping the needles.
Answer D is wrong. Both technicians are correct.

Glossary

ABS Acronym for Antilock Brake System.

Absolute Pressure The zero point from which pressure is measured.

Ackerman Principle The geometric principle used to provide toe out on turns. The ends of the steering arms are angled so that the inside wheel turns more than the outside wheel when a vehicle is making a turn.

Actuator A device that delivers motion in response to an electrical signal.

Adapter The welds under a spring seat to increase the mounting height or fit a seal to the axle.

Adapter Ring A part that is bolted between the clutch cover and the flywheel on some two-plate clutches when the clutch is installed on a flat flywheel.

A/D Converter Abbreviation for Analog-to-Digital Converter.

Additive An ingredient intended to improve a certain characteristic of the material or fluid.

Adjustable Torque Arm A member used to retain axle alignment and, in some cases, control axle torque. Normally one adjustable and one rigid torque arm are used per axle so the axle can be aligned. This rod has means by which it can be extended or retracted for adjustment purposes.

Adjusting Ring A device that is held in the shift signal valve bore by a press fit pin through the valve body housing. When the ring is pushed in by the adjusting tool, the slots on the ring that engage the pin are released.

After-Cooler A device that removes water and oil from the air by a cooling process. The air leaving an after-cooler is saturated with water vapor, which condenses when a drop in temperature occurs.

Air Bag An air-filled device that functions as the spring on axles that utilize air pressure in the suspension system.

Air Brakes A braking system that uses air pressure to actuate the brakes by means of diaphragms, wedges, or cams.

Air-Brake System A system utilizing compressed air to activate the brakes.

Air Compressor An engine-driven mechanism for supplying high pressure air to the truck brake system. There are basically two types of compressors: those designed to work on in-line engines and those that work on V-type engines. The in-line type is mounted forward and is gear driven, while the V-type is mounted toward the firewall and is camshaft driven. With both types the coolant and lubricant are supplied by the truck engine.

Air Conditioning The control of air movement, humidity, and temperature by mechanical means.

Air Dryer A unit that removes moisture.

Air Filter/Regulator Assembly A device that minimizes the possibility of moisture-laden air or impurities entering a system.

Air Hose An air line, such as one between the tractor and trailer, that supplies air for the trailer brakes.

Air-Over-Hydraulic Brakes A brake system utilizing a hydraulic system assisted by an air pressure system.

Air-Over-Hydraulic Intensifier A device that changes the pneumatic air pressure from the treadle brake valve into hydraulic pressure that controls the wheel cylinders.

Air Shifting The process that uses air pressure to engage different range combinations in the transmission's auxiliary section without a mechanical linkage to the driver.

Air Slide Release A release mechanism for a sliding fifth wheel, which is operated from the cab of a tractor by actuating an air control valve. When actuated, the valve energizes an air cylinder, which releases the slide lock and permits positioning of the fifth wheel.

Air Spring An airfilled device that functions as the spring on axles that utilize air pressure in the suspension system.

Air-Spring Suspension A single or multi-axle suspension relying on air bags for springs and weight distribution of axles.

Air Timing The time required for the air to be transmitted to or released from each brake, starting the instant the driver moves the brake pedal.

Altitude-Compensation System An altitude barometric switch and solenoid used to provide better drivability at more than 1,000 feet above sea level.

Ambient Temperature Temperature of the surrounding or prevailing air. Normally, it is considered to be the temperature in the service area where testing is taking place.

Amboid Gear A gear that is similar to the hypoid type with one exception: the axis of the drive pinion gear is located above the centerline axis of the ring gear.

Amp Abbreviation for ampere.

Ampere The unit for measuring electrical current.

Analog Signal A voltage signal that varies within a given range (from high to low, including all points in between).

Analog-to-Digital Converter (A/D Converter) A device that converts analog voltage signals to a digital format; this is located in a section of the processor called the input signal conditioner.

Analog Volt/Ohmmeter (AVOM) A test meter used for checking voltage and resistance. Analog meters should not be used on solid state circuits.

Annulus The outer member of a simple planetary gear set.

Anticorrosion Agent A chemical used to protect metal surfaces from corrosion.

Antifreeze A compound, such as alcohol or glycerin that is added to water to lower its freezing point.

Antilock Brake System (ABS) A computer controlled brake system having a series of sensing devices at each wheel that control braking action to prevent wheel lockup.

Antilock Relay Valve (ARV) In an antilock brake system, the device that usually replaces the standard relay valve used to control the rear-axle service brakes and performs the standard relay function during tractor/trailer operation.

Antirattle Springs Springs that reduce wear between the intermediate plate and the drive pin, and helps to improve clutch release.

Antirust Agent An additive used with lubricating oils to prevent rusting of metal parts when the engine is not in use.

Application Valve A foot-operated brake valve that controls air pressure to the service chambers.

Applied Moment A term meaning a given load has been placed on a frame at a particular point.

Area The total cross section of a frame rail including all applicable elements usually given in square inches.

Armature The rotating component of a (1) starter or other motor, (2) generator, (3) compressor clutch.

Articulating Upper Coupler A bolster plate kingpin arrangement that is not rigidly attached to the trailer, but provides articulation and/or oscillation, (such as a frameless dump) about an axis parallel to the rear axle of the trailer.

Articulation Vertical movement of the front driving or rear axle relative to the frame of the vehicle to which they are attached.

ASE Acronym for Automotive Service Excellence, a trademark of National Institute for Automotive Service Excellence.

Aspect Ratio A tire term calculated by dividing the tire's section height by its section width.

ATA American Trucking Association. Sets standards and influence regulations in the North American truck industry.

Atmospheric Pressure The weight of the air at sea level; 14.696 pounds per square inch (psi) or 101.33 kilopascals (kPa).

Automatic Slack Adjuster The device that automatically adjusts the clearance between the brake linings and the brake drum or rotor. The slack adjuster controls the clearance by sensing the length of the stroke of the push rod for the air brake chamber.

Autoshift Finger The device that engages the shift blocks on the yoke bars that corresponds to the tab on the end of the gearshift lever in manual systems.

Auxiliary Filter A device installed in the oil return line between the oil cooler and the transmission to prevent debris from being flushed into the transmission causing a failure. An auxiliary filter must be installed before the vehicle is placed back in service.

Auxiliary Section The section of a transmission where range shifting occurs, housing the auxiliary drive gear, auxiliary main shaft assembly, auxiliary countershaft, and the synchronizer assembly.

Axis of Rotation The centerline around which a gear or part revolves.

Axle (1) A rod or bar on which wheels turn. (2) A shaft that transmits driving torque to the wheels.

Axle Range Interlock A feature designed to prevent axle shifting when the interaxle differential is locked out, or when lockout is engaged. The basic shift system operates the same as the standard shift system to shift the axle and engage or disengage the lockout.

Axle Seat A suspension component used to support and locate the spring on an axle.

Axle Shims Thin wedges that may be installed under the leaf springs of single axle vehicles to tilt the axle and correct the U-joint operating angles. Wedges are available in a range of sizes to change pinion angles.

Battery Terminal A tapered post or threaded studs on top of the battery case, or infernally threaded provisions on the side of the battery for connecting the cables.

Beam Solid-Mount Suspension A tandem suspension relying on a pivotal mounted beam, with axles attached at the ends for load equalization. The beam is mounted to a solid center pedestal.

Beam Suspension A tandem suspension relying on a pivotally mounted beam, with axles attached at ends for lead equalization. Beam is mounted to center spring.

Bellows A movable cover or seal that is pleated or folded like an accordion to allow for expansion and contraction.

Bending Moment A term implying that when a load is applied to the frame, it will be distributed across a given section of the frame material.

Bias A tire term where belts and plies are laid diagonally or crisscrossing each other.

Bimetallic Two dissimilar metals joined together that have different bending characteristics when subjected to different changes of temperature.

Blade Fuse A type of fuse having two flat male lugs sticking out for insertion in the female box connectors.

Bleed Air Tanks The process of draining condensation from air tanks to increase air capacity and brake efficiency.

Block Diagnosis Chart A troubleshooting chart that lists symptoms, possible causes, and probable remedies in columns.

Blower Fan A fan that pushes or blows air through a ventilation, heater, or air conditioning system.

Bobtail Proportioning Valve A valve that senses when the tractor is bobtailing and automatically reduces the amount of air pressure that can be applied to the tractor's drive axle(s). This reduces braking force on the drive axles, lessening the chance of a spin out on slippery pavement.

Bobtailing A tractor running without a trailer.

Bogie The axle spring, suspension arrangement on the rear of a tandem axle tractor.

Bolster Plate The flat load-bearing surface under the front of a semitrailer, including the kingpin, which rests on the fifth wheel when coupled.

Bolster-Plate Height The height from the ground to the bolster plate when the trailer is level and empty.

Boss A heavy cast section that is used for support, such as the outer race of a bearing or shaft bore.

Bottoming A condition that occurs when (1) The teeth of one gear touch the lowest point between teeth of a mating gear. (2) The bed or frame of the vehicle strikes the axle, such as may be the case of overloading.

Bottom U-Bolt Plate A plate that is located on the bottom side of the spring or axle and is held in place when the U-bolts are tightened to the clamp spring and axle together.

Bracket An attachment used to secure parts to the body or frame.

Brake Control Valve A dual brake valve that releases air from the service reservoirs to the service lines and brake chambers. The valve includes a piston that pushes on diaphragms to open ports; these vent air to service lines in the primary and secondary systems.

Brake Disc A steel disc used in a braking system with a caliper and pads. When the brakes are applied, the pad on each side of the spinning disc is forced against the disc, thus imparting a braking force. This type of brake is very resistant to brake fade.

Brake Drum A cast metal bell-like cylinder attached to the wheel that is used to house the brake shoes and provide a friction surface for stopping a vehicle.

Brake Fade A condition that occurs when friction surfaces become hot enough to cause the coefficient of friction to drop to a point where the application of severe pedal pressure results in little actual braking.

Brake Lining Friction material used to line brake shoes or brake pads. It withstands high temperatures and pressure. The molded material is either riveted or bonded to the brake shoe, with a suitable coefficient of friction for stopping a vehicle.

Brake Pad The friction lining and plate assembly that is forced against the rotor to cause braking action in a disc brake system.

Brake Shoe The curved metal part, faced with brake lining, which is forced against the brake drum to produce braking action.

Brake-Shoe Rollers A hardware part that attaches to the web of the brake shoes by means of roller retainers. The rollers, in turn, ride on the end of an S-cam.

Brake System The vehicle system that slows or stops a vehicle. A combination of brakes and a control system.

British Thermal Unit (Btu) A measure of heat quantity equal to the amount of heat required to raise 1 pound of water 1°F.

Broken-Back Driveshaft A term often used for nonparallel driveshaft.

Btu Acronym for British Thermal Unit.

Bump Steer Erratic steering caused from rolling over bumps, cornering, or heavy braking. Same as orbital steer and roll steer.

CAA Acronym for Clean Air Act.

Caliper A disc brake component that changes hydraulic pressure into mechanical force and uses that force to press the brake pads against the rotor and stop the vehicle. Calipers come in three basic types: fixed, floating, and sliding, and can have one or more pistons.

Camber The attitude of a wheel and tire assembly when viewed from the front of a car. If it leans outward, away from the car at the top, the wheel is said to have positive camber. If it leans inward, it is said to have negative camber.

Cam Brakes Brake shoes that are actuated by an S-type camshaft that forces them into a drum.

Cartridge Fuse A type of fuse having a strip of low melting point metal enclosed in a glass tube. If an excessive current flows through the circuit, the fuse element melts at the narrow portion, opening the circuit and preventing damage.

Caster The angle formed between the kingpin axis and a vertical axis as viewed from the side of the vehicle. Caster is considered positive when the top of the kingpin axis is behind the vertical axis.

Cavitation Erosion caused by vapor-bubble collapse.

Center of Gravity The point around which the weight of a truck is evenly distributed; the point of balance.

Ceramic Fuse A fuse found in some import vehicles that has a ceramic insulator with a conductive metal strip along one side.

CFC Acronym for chlorofluorocarbon.

Charging System A system consisting of the battery, alternator, voltage regulator, associated wiring, and the electrical loads of a vehicle. The purpose of the system is to recharge the battery whenever necessary and to provide the current required to power the electrical components.

Charging Circuit The alternator (or generator) and associated circuit used to keep the battery charged and to furnish power to the vehicle's electrical systems when the engine is running.

Check Valve A valve that allows air to flow in one direction only. It is a federal requirement to have a check valve between the wet and dry air tanks.

Chlorofluorocarbon (CFC) A compound used in older A/C refrigerant that damages the ozone layer.

Circuit The complete path of an electrical current, including the generating device. When the path is unbroken, the circuit is closed and current flows. When the circuit continuity is broken, the circuit is open and current flow stops.

Clean Air Act (CAA) Federal regulations, passed in 1992, that have resulted in major changes in air conditioning systems.

Climbing A gear problem caused by excessive wear in gears, bearings, and shafts whereby the gears move sufficiently apart to cause the apex (or point) of the teeth on one gear to climb over the apex of the teeth on another gear with which it is meshed.

Clutch A device for connecting and disconnecting the engine from the transmission or for a similar purpose in other units.

Clutch Brake A circular disc with outboard friction surfaces mounted on the transmission input shaft between the release bearing and the transmission. Its purpose is to slow or stop the transmission input shaft from rotating in order to allow gears to be engaged without clashing or grinding.

Clutch Housing A component that surrounds and protects the clutch and connects the transmission case to the vehicle's engine.

Clutch Pack An assembly of normal clutch plates, friction discs, and one very thick plate known as the pressure plate. The pressure plate has tabs around the outside diameter to mate with the channel in the clutch drum.

COE Acronym for cab-over-engine.

Coefficient of Friction A measurement of the amount of friction developed between two objects in physical contact when one of the objects is drawn across the other.

Coil Springs Spring steel spirals that are mounted on control arms or axles to absorb road shock.

Combination A truck coupled to one or more trailers.

Compression Applying pressure to a spring or any springy substance, thus causing it to reduce its length in the direction of the compressing force.

Compressor (1) A mechanical device that increases pressure within a container by pumping air into it. (2) That component of an air conditioning system that compresses low temperature/pressure refrigerant vapor.

Condensation The process by which gas (or vapor) changes to a liquid.

Condenser A component in an air conditioning system used to cool a refrigerant below its boiling point, causing it to change from a vapor to a liquid.

Conductor Any material that permits the electrical current to flow.

Constant-Rate Springs Leaf-type spring assemblies that have a constant rate of deflection.

Control Arm The main link between the vehicle's frame and the wheels that acts as a hinge to allow wheel action up and down independent of the chassis.

Controlled Traction A type of differential that uses a friction plate assembly to transfer drive torque from the vehicle's slipping wheel to the one wheel that has good traction or surface bite.

Converter Dolly An axle, frame, drawbar, and fifth wheel arrangement that converts a semitrailer into a full trailer.

Coolant Liquid that circulates in an engine cooling system.

Coolant Heater A component used to aid engine starting and reduce the wear caused by cold starting.

Coolant Hydrometer A tester designed to measure coolant specific gravity and determine the amount of antifreeze in the coolant.

Cooling System Complete system for circulating coolant.

Coupling Point The point at which the turbine is turning at the same speed as the impeller.

Crankcase The housing within which the crankshaft and many other parts of the engine operate.

Cranking Circuit The starter and its associated circuit, including battery, relay (solenoid), ignition switch, neutral start switch (on vehicles with automatic transmission), cables, and wires.

Cross Groove Joint disc-shaped type of inner CV joint that uses balls and V-shaped grooves on the inner and outer races to accommodate the plunging motion of the half-shaft. The joint usually bolts to a transaxle stub flange; same as disc-type joint.

Cross-Tube A system that transfers the steering motion to the opposite, passenger side steering knuckle. It links the two steering knuckles together and forces them to act in unison.

C-Train A combination of two or more trailers in which the dolly is connected to the trailer by means of two pintle hook or coupler drawbar connections. The resulting connection has one pivot point.

CVSA An acronym for the Commercial Vehicle Safety Alliance, which is a nonprofit organization promoting safety.

Cycling (1) Repeated on-off action of the air conditioner compressor. (2) Repeated electrical discharge/recharge cycling that can cause the battery positive plate material to break away from its grids and fall into the sediment chambers at the base of the battery case.

Dampen To slow or reduce oscillations or movement.

Dampened Discs Clutch discs that have dampening springs incorporated into the disc hub. When engine torque is first transmitted to the disc, the plate rotates on the hub, compressing the springs. This action absorbs the shocks and torsional vibration caused by today's low rpm, high torque, engines.

Dash Control Valves A variety of hand operated valves located on the dash. They include parking brake valves, tractor protection valves, and differential lock.

Data Links Circuits through which computers communicate with other electronic devices such as control panels, modules, some sensors, or other computers in the form of digital signals.

Dead Axle Nonlive or dead axles are often mounted in lifting suspensions. They hold the axle off the road when the vehicle is traveling empty, and put it on the road when a load is being carried. They are also used as air suspension third axles on heavy straight trucks and are used extensively in eastern states with high axle weight laws. An axle that does not rotate but merely forms a base on which to attach the wheels.

Deadline To take a vehicle out of service.

Deburring To remove sharp edges from a cut.

Dedicated-Contract Carriage Trucking operations set up and run according to a specific shipper's needs. In addition to transportation, they often provide other services such as warehousing and logistics planning.

Deflection Bending or moving to a new position as the result of an external force.

Department of Transportation (DOT) A government agency that establishes vehicle standards.

Detergent Additive An additive that helps keep metal surfaces clean and prevents deposits. These additives suspend particles of carbon and oxidized oil in the oil.

DER Acronym for Department of Environmental Resources.

Diagnostic Flow Chart A chart that provides a systematic approach to the electrical system and component troubleshooting and repair. They are found in service manuals and are vehicle make and model specific.

Dial Caliper A measuring instrument capable of taking inside, outside, depth, and step measurements.

Differential A gear assembly that transmits power from the driveshaft to the wheels and allows two opposite wheels to turn at different speeds for cornering and traction.

Differential-Carrier Assembly An assembly that controls the drive axle operation.

Differential Lock A toggle or push-pull type air switch that locks together the rear axles of a tractor so they pull as one for off-the-road operation.

Digital Binary Signal A signal that has only two values; on and off.

Digital Volt/Ohmmeter (DVOM) A type of test meter recommended by most manufacturers for use on solid state circuits.

Diode The simplest semiconductor device formed by joining P-type semiconductor material with N-type semiconductor material. A diode allows current to flow in one direction, but not in the opposite direction.

Direct Drive The gearing of a transmission so that one revolution of the engine produces one revolution of the transmission's output shaft.

Disc Brake A steel rotor used in a braking system with a caliper and pads. When the brakes are applied, the pad on each side of the spinning disc is forced against the disc, thus imparting a braking force. This type of brake is very resistant to brake fade. A type of brake that generates stopping power by the application of pads against a rotating disc (rotor).

Dispatch Sheet A form used to keep track of dates when the work is to be completed. Some dispatch sheets follow the job through each step of the servicing process.

Dog Tracking Off-center tracking of the rear wheels as related to the front wheels.

DOT Acronym for Department of Transportation.

Downshift Control The selection of a lower range to match driving conditions encountered or expected to be encountered. Learning to take advantage of a downshift gives better control on slick or icy roads and on steep downgrades. Downshifting to lower ranges increases engine braking.

Double Reduction Axle An axle that uses two gear sets for greater overall gear reduction and peak torque development. This design is favored for severe service applications, such as dump trucks, cement mixers, and other heavy haulers.

Drag Link A connecting rod or link between the steering gear, pitman arm, and the steering linkage.

Drawbar Capacity The maximum, horizontal pulling force that can be safely applied to a coupling device.

Driven Gear A gear that is driven or forced to turn by a drive gear, by a shaft, or by some other device.

Drive or Driving Gear A gear that drives another gear or causes another gear to turn.

Driveline The propeller or driveshaft, universal joints, and so forth, that links the transmission output to the axle pinion gear shaft.

Driveline Angle The alignment of the transmission output shaft, driveshaft, and rear-axle pinion centerline.

Driveshaft An assembly of one or two universal joints connected to a shaft or tube; used to transmit power from the transmission to the differential.

Drivetrain An assembly that includes all power transmitting components from the rear of the engine to the wheels, including clutch/torque converter, transmission, driveline, and front and rear driving axles.

Driver Controlled Main Differential Lock A type of axle assembly has greater flexibility over the standard type of single reduction axle because it provides equal amounts of driveline torque to each driving wheel, regardless of changing road conditions. This design also provides the necessary differential action to the road wheels when the truck is turning a corner.

Driver's Manual A publication that contains information needed by the driver to understand, operate, and care for the vehicle and its components.

Drum Brake A type of brake system in which stopping friction is created by the shoes pressing against the inside of the rotating drum.

Dual-Hydraulic Braking System A brake system consisting of a tandem, or double action master cylinder, which is basically two master cylinders usually formed by aligning two separate pistons and fluid reservoirs into a single cylinder.

ECU Acronym for electronic control unit.

Eddy Current A small circular current produced inside a metal core in the armature of a starter motor. Eddy currents produce heat and are reduced by using a laminated core.

Electricity The movement of electrons from one place to another.

Electric Retarder Electromagnets mounted in a steel frame. Energizing the retarder causes the electromagnets to exert a dragging force on the rotors in the frame and this drag force is transmitted directly to the driveshaft.

Electromotive Force (EMF) The force that moves electrons between atoms. This force is the pressure that exists between the positive and negative points (the electrical imbalance). This force is measured in units called volts.

Electronically Programmable Memory (EPROM) Computer memory that permits adaptation of the ECU to various standard mechanically controlled functions.

Electronic Control Unit (ECU) The brain of the vehicle.

Electronics The technology of controlling electricity.

Electrons Negatively charged particles orbiting around every nucleus.

Elliot Axle A solid bar front axle on which the ends span the steering knuckle.

EMF Acronym for electromotive force.

End Yoke The component connected to the output shaft of the transmission to transfer engine torque to the driveshaft.

Engine Brake Any number of devices that convert the role of the vehicle engine into that of a power absorbing retarding mechanism.

Environmental Protection Agency An agency of the United States government charged with the responsibilities of protecting the environment and enforcing the Clean Air Act (CAA) of 1990.

EPA Acronym for the Environmental Protection Agency.

EPROM Acronym for Electronically Programmable Memory.

Equalizer A suspension device used to transfer and maintain equal load distribution between two or more axles of a suspension. Formerly called a rocker beam.

Equalizer Bracket A bracket for mounting the equalizer beam of a multiple axle spring suspension to a truck or trailer frame while allowing for the beam's pivotal movement. Normally there are three basic types: flange-mount, straddle-mount, and under- or side-mount.

Evaporator A component in an air conditioning system used to remove heat from the air passing through it.

Exhaust Brake A sliding restriction valve in the exhaust system located between the manifold and the muffler. A slide mechanism that restricts the exhaust flow, causing exhaust back pressure to build up in the engine's cylinders and effect retarding.

External Housing Damper A counterweight attached to an arm on the rear of the transmission extension housing and designed to dampen unwanted driveline or powertrain vibrations.

Extra Capacity A term that generally refers to (1) A coupling device that has strength capability greater than standard. (2) An oversized tank or reservoir for a fluid or vapor.

False Brinelling The polishing of a surface that is not damaged.

Fanning the Brakes Applying and releasing the brakes in rapid succession on a long downgrade.

Fatigue Failures The progressive destruction of a shaft or gear teeth material usually caused by overloading.

Fault Code A code that is recorded into the computer's memory. A fault code can be read by plugging a special break-out box tester into the computer.

Federal Motor Vehicle Safety Standard (FMVSS) A federal standard that specifies that all vehicles in the United States be assigned a Vehicle Identification Number (VIN).

Federal Motor Vehicle Safety Standard No. 121 (FMVSS 121) A federal standard that governs standards in air brake systems. Most highway trucks and trailers built since 1976 are required to meet FMVSS standards that are routinely updated as technology advances.

FHWA Acronym for Federal Highway Administration.

Fiber Composite Springs Springs that are made of fiberglass, laminated, and bonded together by tough polyester resins.

Fifth Wheel A coupling device mounted on a truck and used to connect a semitrailer. It acts as a hinge point to allow changes in direction of travel between the tractor and the semitrailer.

Fifth-Wheel Height The distance from the ground to the top of the fifth wheel when it is level and parallel with the ground. It can also refer to the height from the tractor frame to the top of the fifth wheel. The latter definition applies to data given in fifth wheel literature.

Fifth-Wheel Top Plate The portion of the fifth wheel assembly that contacts the trailer bolster plate and houses the locking mechanism that connects to the kingpin.

Final Drive The last reduction gear set of a truck.

Fixed Value Resistor An electrical device that is designed to have only one resistance rating, which should not change, for controlling voltage.

Flammable Any material that will easily catch fire or explode.

Flare To spread gradually outward in a bell shape.

Flex Disc A term often used for flexplate.

Flexplate A component used to mount the torque converter to the crankshaft. The flex plate is positioned between the engine crankshaft and the T/C. The purpose of the flex plate is to transfer crankshaft rotation to the shell of the torque converter assembly.

FMVSS Acronym for Federal Motor Vehicle Safety Standard.

FMVSS No 121 Acronym for Federal Motor Vehicle Safety Standard No 121.

Foot Valve A foot-operated brake valve that controls air pressure to the service chambers.

Foot-Pound An English unit of measurement for torque. One foot-pound is the torque obtained by a force of 1 pound applied to a foot long wrench handle.

Forged Journal Cross Part of a universal joint.

Frame Width The measurement across the outside of the frame rails of a tractor, truck, or trailer.

Franchised Dealership A dealer that has signed a contract with a particular manufacturer to sell and service a particular line of vehicles.

Fretting A result of vibration that the bearing outer race can pick up from the machining pattern.

Friction-Plate Assembly An assembly consisting of a multiple disc clutch that is designed to slip when a predetermined torque value is reached.

Front-Axle Limiting Valve A valve that reduces pressure to the front service chambers, thus eliminating front wheel lockup on wet or icy pavements.

Front Hanger A bracket for mounting the front of the truck or trailer suspensions to the frame. Made to accommodate the end of the spring on spring suspensions. There are four basic types: flange-mount, straddle-mount, under-mount, and side-mount.

Full Trailer A trailer that does not transfer load to the towing vehicle. It employs a tow bar coupled to a swiveling or steerable running gear assembly at the front of the trailer.

Fully Floating Axles An axle configuration whereby the axle half-shafts transmit only driving torque to the wheels and not bending and torsional loads that are characteristic of the semi-floating axle.

Fully-Oscillating Fifth Wheel A fifth wheel with universal articulation.

Fusible Link A term often used for fuse link.

Fuse Link A short length of smaller gauge wire installed in a conductor, usually close to the power source.

GCW Acronym for gross combination weight.

Gear A disk-like wheel with external or internal teeth that serves to transmit or change motion.

Gear Pitch The number of teeth per given unit of pitch diameter, an important factor in gear design and operation.

Gladhand The connectors between tractor and trailer air lines.

Gross Combination Weight (GCW) The total weight of a fully quipped vehicle including payload, fuel, and driver.

Gross Trailer Weight (GTW) The sum of the weight of an empty trailer and its payload.

Gross Vehicle Weight (GVW) The total weight of a fully equipped vehicle and its payload.

Ground The negatively charged side of a circuit. A ground can be a wire, the negative side of the battery, or the vehicle chassis.

Grounded Circuit A shorted circuit that causes current to return to the battery before it has reached its intended destination.

GTW Acronym for gross trailer weight.

GVW Acronym for gross vehicle weight.

Halogen Light A lamp having a small quartz/glass bulb that contains a fuel filament surrounded by halogen gas. It is contained within a larger metal reflector and lens element.

Hand Valve (1) A valve mounted on the steering column or dash, used by the driver to apply the trailer brakes independently of the tractor brakes. (2) A hand operated valve used to control the flow of fluid or vapor.

Harness and Harness Connectors The organization of the vehicle's electrical system providing an orderly and convenient starting point for tracking and testing circuits.

Hazardous Materials Any substance that is flammable, explosive, or is known to produce adverse health effects in people or the environment that are exposed to the material during its use.

Heads-Up Display (HUD) A technology used in some vehicles that superimposes data on the driver's normal field of vision. The operator can view the information, which appears to "float" just above the hood at a range near the front of a conventional tractor or truck. This allows the driver to monitor conditions such as limited road speed without interrupting his normal view of traffic.

Heater Control Valve A valve that controls the flow of coolant into the heater core from the engine.

Heat Exchanger A device used to transfer heat, such as a radiator or condenser.

Heavy-Duty Truck A truck that has a GVW of 26,001 pounds or more.

Helper Spring An additional spring device that permits greater load on an axle.

High CG Load Any application in which the load center of gravity (CG) of the trailer exceeds 40 inches (102 centimeters) above the top of the fifth wheel.

High-Resistant Circuits Circuits that have an increase in circuit resistance, with a corresponding decrease in current.

High-Strength Steel A low-alloy steel that is much stronger than hot-rolled or cold-rolled sheet steels that normally are used in the manufacture of car body frames.

Hinged Pawl Switch The simplest type of switch; one that makes or breaks the current of a single conductor.

HUD Abbreviation for heads up display.

Hydraulic Brakes Brakes that are actuated by a hydraulic system.

Hydraulic Brake System A system utilizing the properties of fluids under pressure to activate the brakes.

Hydrometer A tester designed to measure the specific gravity of a liquid.

I-Beam Axle An axle designed to give great strength at reasonable weight. The cross section of the axle resembles the letter "I."

ICC Check Valve A valve that allows air to flow in one direction only. It is a federal requirement to have a check valve between the wet and dry air tanks.

Inboard Toward the centerline of the vehicle.

In-Line Fuse A fuse that is in series with the circuit in a small plastic fuse holder, not in the fuse box or panel. It is used, when necessary, as a protection device for a portion of the circuit even though the entire circuit may be protected by a fuse in the fuse box or panel.

In-Phase The in-line relationship between the forward coupling shaft yoke and the driveshaft slip yoke of a two-piece driveline.

Input Retarder A device located between the torque converter housing and the main housing designed primarily for over-the-road operations. The device employs a "paddle wheel" type design with a vaned rotor mounted between stator vanes in the retarder housing.

Installation Templates Drawings supplied by some vehicle manufacturers to allow the technician to correctly install an accessory. The templates available can be used to check clearances or to ease installation.

Insulator A material, such as rubber or glass, that offers high electrical resistance.

Integrated Circuit A component containing diodes, transistors, resistors, capacitors, and other electronic components mounted on a single piece of material and capable to perform numerous functions.

Jacobs Engine Brake A hydraulically operated device that converts a power producing diesel engine into a power-absorbing retarder mechanism by altering the engine's exhaust valve opening time used to slow the vehicle.

Jake Brake A slang term for Jacobs engine brake.

Jumper Wire A wire used to temporarily bypass a circuit or components for electrical testing. A jumper wire consists of a length of wire with an alligator clip at each end.

Jumpout A condition that occurs when a fully engaged gear and sliding clutch are forced out of engagement.

Jump Start The procedure used when it becomes necessary to use a booster battery to start a vehicle having a discharged battery.

Kinetic Energy Energy in motion.

Kingpin (1) The pin mounted through the center of the trailer upper coupler (bolster plate) that mates with the fifth wheel locks, securing the trailer to the fifth wheel. The configuration is controlled by industry standards. (2) A pin or shaft on which the steering spindle rotates.

Kompensator Fifth Wheel A fifth wheel with fore-and-aft and side-to-side articulation.

Landing Gear The retractable supports for a semitrailer to keep the trailer level when the tractor is detached from it.

Laser Beam Alignment System A two- or four-wheel alignment system using wheel-mounted instruments to project a laser beam to measure toe, caster, and camber.

Lateral Runout The wobble or side-to-side movement of a rotating wheel or of a rotating wheel and tire assembly.

Leaf Springs Strips of steel connected to the chassis and axle to isolate the vehicle from road shock.

Less Than Truckload (LTL) Partial loads from the networks of consolidation centers and satellite terminals.

Light Beam Alignment System An alignment system using wheel-mounted instruments to project light beams onto charts and scales to measure toe, caster, and camber, and note the results of alignment adjustments.

Limited-Slip Differential A differential that utilizes a clutch device to deliver power to either rear wheel when the opposite wheel is spinning.

Linkage A system of rods and levers used to transmit motion or force.

Live Axle An axle on which the wheels are firmly affixed. The axle drives the wheels.

Live-Beam Axle A nonindependent suspension in which the axle moves with the wheels.

Load Proportioning Valve (LPV) A valve used to redistribute application pressure to front and rear brakes based on vehicle loads. This is a load- or height-sensing valve that senses the vehicle load and proportions the braking between front and rear brakes.

Lockstrap A manual clutch adjustment mechanism that allows for the adjustment of free travel.

Lock-up Torque Converter A torque converter that eliminates the 10 percent slip that takes place between the impeller and turbine at the coupling stage of operation.

Longitudinal Leaf Spring A leaf spring that is mounted so it is parallel to the length of the vehicle.

Low-Maintenance Battery A conventionally vented, lead/acid battery, requiring normal periodic maintenance.

LTL Acronym for less than truckload.

Maintenance-Free Battery A battery that does not require the addition of water during normal service life.

Maintenance Manual A publication containing routine maintenance procedures and intervals for vehicle components and systems.

Main Transmission A transmission consisting of an input shaft, floating main shaft assembly and main drive gears, two countershaft assemblies, and reverse idler gears.

Manual Slide Release The release mechanism for a sliding fifth wheel, which is operated by hand.

Metering Valve A valve used on vehicles equipped with front disc and rear drum brakes. It improves braking balance during light brake applications by preventing application of the front disc brakes until pressure is built up in the hydraulic system.

Moisture Ejector A valve mounted to the bottom or side of the supply and service reservoirs that collect water and expel it every time the air pressure fluctuates.

Mounting Bracket That portion of the fifth wheel assembly that connects the fifth wheel top plate to the tractor frame or fifth wheel mounting system.

Multiaxle Suspension A suspension consisting of more than three axles.

Multiple-Disc Clutch A clutch having a large drum-shaped housing that can be either a separate casting or part of the existing transmission housing.

NATEF Acronym for National Automotive Technicians Education Foundation.

National Automotive Technicians Education Foundation (NATEF) A foundation having a program of certifying secondary and post secondary automotive and heavy-duty truck training programs.

National Institute for Automotive Service Excellence (ASE) A nonprofit organization that has an established certification program for automotive, heavy-duty truck, auto body repair, engine machine shop technicians, and parts specialists.

Needlenose Pliers This tool has long tapered jaws for grasping small parts or for reaching into tight spots. Many needlenose pliers also have cutting edges and a wire stripper.

NIASE Acronym for National Institute for Automotive Service Excellence, now abbreviated ASE.

NIOSH Acronym for National Institute for Occupation Safety and Health.

NLGI Acronym for National Lubricating Grease Institute.

NHTSA Acronym for National Highway Traffic Safety Administration.

Nonlive Axle Nonlive or dead axles are often mounted in lifting suspensions. They hold the axle off the road when the vehicle is traveling empty, and put it on the road when a load is being carried. They are also used as air suspension third axles on heavy straight trucks and are used extensively in eastern states with high axle weight laws.

Nonparallel Driveshaft A type of driveshaft installation whereby the working angles of the joints of a given shaft are equal; however the companion flanges and/or yokes are not parallel.

Nonpolarized Gladhand A gladhand that can be connected to either service or emergency gladhand.

Nose The front of a semitrailer.

No-tilt Convertible Fifth A fifth wheel with fore/aft articulation that can be locked out to produce a rigid top plate for applications that have either rigid and/or articulating upper couplers.

OEM Acronym for original equipment manufacturer.

Off-road Any terrain not considered part of the highway system.

Ohm A unit of measured electrical resistance.

Ohm's Law The basic law of electricity stating that in any electrical circuit, current, resistance, and pressure work together in a mathematical relationship.

On-road With reference to paved or smooth-graded surface terrain on which a vehicle will operate, generally considered to be part of the public highway system.

OOS Acronym for out-of-service (deadlined).

Open Circuit An electrical circuit whose path has been interrupted or broken either accidentally (a broken wire) or intentionally (a switch turned off).

Operational Control Valve A valve used to control the flow of compressed air through the brake system.

Oscillation The rotational movement in either fore/aft or side-to-side direction about a pivot point. Generally refers to fifth wheel designs in which fore/aft and side-to-side articulation are provided.

OSHA Acronym for Occupational Safety and Health Administration.

Out-of-Phase A condition of the universal joint that acts somewhat like one person snapping a rope held by a person at the opposite end. The result is a violent reaction at the opposite end. If both were to snap the rope at the same time, the resulting waves would cancel each other and neither would feel the reaction.

Out-of-Round A wheel or tire defect in which the wheel or tire is not round.

Output Driver An electronic on/off switch that the computer uses to control the ground circuit of a specific actuator. Output drivers are located in the processor along with the input conditioners, microprocessor, and memory.

Output Yoke The component that serves as a connecting link, transferring torque from the transmission's output shaft through the vehicle's driveline to the rear axle.

Oval Eccentric or egg-shaped.

Overall Ratio The ratio of the lowest to the highest forward gear in the transmission.

Overdrive The gearing of a transmission so that in its highest gear one revolution of the engine produces more than one revolution of the transmission's output shaft.

Overrunning Clutch A clutch mechanism that transmits power in one direction only.

Overspeed Governor A governor that shuts off the fuel or stops the engine when excessive speed is reached.

Oxidation Inhibitor (1) An additive used with lubricating oils to keep oil from oxidizing even at very high temperatures. (2) An additive for gasoline to reduce the chemicals in gasoline that react with oxygen.

Pad A disc brake lining and metal back riveted, molded, or bonded together.

Parallel Circuit An electrical circuit that provides two or more paths for the current to flow. Each path has separate resistors and operates independently from the other parallel paths. In a parallel circuit, amperage can flow through more than one resistor at a time.

Parallel-Joint Type A type of driveshaft installation whereby all companion flanges and/or yokes in the complete driveline are parallel to each other with the working angles of the joints of a given shaft being equal and opposite.

Parking Brake A mechanically applied brake used to prevent a parked vehicle's movement.

Parts Requisition A form that is used to order new parts, on which the technician writes the names of what part(s) are needed along with the vehicle's VIN or company's identification folder.

Payload The weight of the cargo carried by a truck, not including the weight of the body.

Pipe or Angle Brace Extrusions between opposite hangers on a spring or air-type suspension.

Pitman Arm A steering linkage component that connects the steering gear to the linkage at the left end of the center link.

Pitting Surface irregularities resulting from corrosion.

Planetary Drive A planetary gear reduction set where the sun gear is the drive and the planetary carrier is the output.

Planetary Gear Set A system of gearing that is somewhat like the solar system. A pinion is surrounded by an internal ring gear and planet gears are in mesh between the ring gear and pinion around which all revolve.

Planetary Pinion Gears Small gears fitted into a framework called the planetary carrier.

Plies The layers of rubber-impregnated fabric that make up the body of a tire.

PM Acronym for preventive maintenance.

PMI Acronym for preventive maintenance inspections. There are three recognized categories, which are: The "A" PM service, which is the simplest, the "B" inspection, which is described as a major inspection, and the "C" inspection, which is a major inspection plus adjustments and minor repairs.

Pogo Stick The air and electrical line support rod mounted behind the cab to keep the lines from dragging between the tractor and trailer.

Polarity The particular state, either positive or negative, with reference to the two poles or to electrification.

Pole The number of input circuits made by an electrical switch.

Pounds per Square Inch (psi) A unit of standard measurement for pressure.

Power A measure of work being done factored with time.

Power Flow The flow of power from the input shaft through one or more sets of gears, or through an automatic transmission to the output shaft.

Power Steering A steering system utilizing hydraulic pressure to reduce the turning effort required of the operator.

Power Synchronizer A device to speed up the rotation of the main section gearing for smoother automatic downshifts and to slow down the rotation of the main section gearing for smoother automatic upshifts.

Powertrain Power delivery components from the engine through to the wheels.

Pressure The amount of force applied to a defined area measured in pounds per square inch (psi) English or kilopascals (kPa) metric.

Pressure Differential The difference in pressure between any two points of a system or a component.

Pressure Relief Valve (1) A valve located on the wet tank, usually preset at 150 psi (1,034 kPa). Limits system pressure if the compressor or governor unloader valve malfunctions. (2) A valve located on the rear head of an air conditioning compressor or pressure vessel that opens if an excessive system pressure is exceeded.

Printed Circuit Board An electronic circuit board made of thin nonconductive plastic-like material onto which conductive metal, such as copper, has been deposited. Parts of the metal are then etched away by an acid, leaving metal lines that form the conductors for the various circuits on the board. A printed circuit board can hold many complex circuits in a very small area.

Programmable Read Only Memory (PROM) An electronic component that contains program information specific to different vehicle model calibrations.

PROM Acronym for Programmable Read Only Memory.

Proportioning Valve A valve used on vehicles equipped with front disc and rear drum brakes. It is installed in the lines to the rear drum brakes, and in a split system, below the pressure differential valve. By reducing pressure to the rear drum brakes, the valve helps to prevent premature lockup during severe brake application and provides better braking balance.

psi Acronym for pounds per square inch.

Pull Circuit A circuit that brings the cab from a fully tilted position up and over the center.

Pull-Type Clutch A type of clutch that does not push the release bearing toward the engine; instead, it pulls the release bearing toward the transmission.

Pump/Impeller Assembly The input (drive) member that receives power from the engine.

Push Circuit A circuit that raises the cab from the lowered position to the desired tilt position.

Push-Type Clutch A type of clutch in which the release bearing is not attached to the clutch cover.

Quick Release Valve A device used to exhaust air as close as possible to the service chambers or spring brakes.

Radial A tire design having cord materials running in a direction from the center point of the tire, usually from bead to bead.

Radial Load A load that is applied at 90 degrees to an axis of rotation.

RAM Acronym for random access memory.

Ram Air Air that is forced into the engine or passenger compartment by the forward motion of the vehicle.

Random Access Memory (RAM) The memory used during computer operation to store temporary information. The microprocessor can write, read, and erase information from RAM in any order, which is why it is called random.

Range Shift Cylinder A component located in the auxiliary section of the transmission. This component, when directed by air pressure via low and high ports, shifts between high and low range of gears.

Range Shift Lever A lever located on the shift knob allows the driver to select low or high gear range.

Rated Capacity The maximum, recommended safe load that can be sustained by a component or an assembly without permanent damage.

Ratio Valve A valve used on the front or steering axle of a heavy-duty truck to limit the brake application pressure to the actuators during normal service braking.

RCRA Acronym for Resource Conservation and Recovery Act.

Reactivity The characteristic of a material that causes it to react with air, heat, water, or other materials.

Read Only Memory (ROM) Optically or magnetically retained memory used in microcomputers to store information permanently.

Rear Hanger A bracket for mounting a truck or trailer suspension to the frame. There are four types: flange-mount, straddle-mount, under-mount, and side-mount.

Recall Bulletin A notification bulletin that applies to service work or replacement of parts in connection with a recall notice.

Reference Voltage The voltage supplied to a sensor by a vehicle computer.

Relay An electric switch that allows a small current to control a much larger one. It consists of a control circuit and a power circuit.

Relay/Quick Release Valve A pneumatic valve used on air-brake equipped trucks. It is used to speed the application and release of air to the service chambers.

Refrigerant A fluid capable of vaporizing at a low temperature used in A/C and reefer systems.

Refrigerant Management Center Equipment designed to recover, recycle, and recharge an air conditioning system.

Release Bearing A unit used to disengage a clutch that mounts on the transmission input shaft but does not rotate with it.

Reserve Capacity Rating The ability of a battery to sustain a minimum vehicle electrical load in the event of a charging system failure.

Resistance The opposition to current flow in an electrical circuit.

Resist Bend Moment A measurement of frame rail strength calculated by multiplying the section modulus of the rail by the yield strength of the material. This term is universally used in evaluating frame rail strength.

Resource Conservation and Recovery Act (RCRA) A law that states that after using a hazardous material, it must be properly stored until an approved hazardous waste hauler arrives to take it to the disposal site.

Reverse Elliot Axle A solid-beam front axle on which the steering knuckles span the axle ends.

Revolutions per Minute (rpm) The number of complete turns a member makes in one minute.

Right-to-Know Law A law passed by the federal government and administered by the Occupational Safety and Health Administration (OSHA) that requires any company that uses or produces hazardous chemicals or substances to inform its employees, customers, and vendors of any potential hazards that may exist in the workplace as a result of using the products.

Rigid Disc A steel clutch plate to which friction linings, or facings, are bonded or riveted.

Rigid Fifth Wheel A coupling plate that is fixed rigidly to a frame. This fifth wheel has no articulation or oscillation. It is used in applications where the articulation is provided by other means, such as an articulating upper coupler of a frame-less dump.

Rigid-Torque Arm A member used to retain axle alignment and, in some cases, to control axle torque. Normally, one adjustable and one rigid arm are used per axle so the axle can be aligned.

Ring Gear (1) The cranking gear of a flywheel. (2) A large circular gear such as that found in a final drive assembly.

Rocker Beam A suspension device used to transfer and maintain equal load distribution between two or more axles of a suspension.

Roll Axis The theoretical line that joins the roll center of the front and rear axles.

Roller Clutch A clutch designed with a movable inner race, rollers, accordion (apply) springs, and outer race. Around the inside diameter of the outer race are several cam-shaped pockets. The clutch assembly rollers and accordion springs are located in these pockets.

Rollers A hardware part that attaches to the web of the brake shoes by means of roller retainers. The rollers, in turn, ride on the end of an S-cam.

ROM Acronym for read-only memory.

Rotary Oil Flow A condition caused by the centrifugal force applied to the fluid as the converter rotates around its axis.

Rotation A term used to describe the fact that a gear, shaft, or other device is turning.

rpm Acronym for revolutions per minute.

Rotor (1) A part of the alternator that provides the magnetic fields necessary to create a current flow. (2) The rotating member of an assembly.

Runout A deviation of the specified normal travel of an object. The amount of deviation or wobble a shaft or wheel has as it rotates. Runout is measured with a dial indicator.

Safety Factor (SF) (1) The amount of load that can safely be absorbed by and through the vehicle chassis frame members. (2) The difference between the stated and rated limits of a product, such as a grinding disk.

Screw Pitch Gauge A gauge used to check the threads per inch of a nut or bolt.

Secondary Lock The component of a fifth-wheel locking mechanism that backs up the primary lock. The secondary lock can only be engaged if the primary lock is properly engaged.

Section Height The tread center to bead plane on a tire.

Section Width The measurement on a tire from sidewall to sidewall.

Self-Adjusting Clutch A clutch that automatically takes up the slack between the pressure plate and clutch disc as wear occurs.

Semiconductor A solid state device that can function as either a conductor or an insulator, depending on how its structure is arranged.

Semifloating Axle An axle shaft in which drive torque from the differential is transferred directly to the wheels. A single bearing assembly, located at the outer end of the axle, is used to support the axle half-shaft.

Semioscillating A term that generally describes a fifth wheel type that oscillates or articulates about an axis perpendicular to the vehicle centerline.

Semitrailer A load-carrying vehicle equipped with one or more axles and constructed so that its front end is supported on the fifth wheel of the truck tractor that pulls it.

Sensing Voltage The voltage that allows the regulator to sense and monitor the battery voltage level.

Sensor An electronic device used to signal vehicle computers.

Series Circuit An electrical circuit with a voltage source and a single path for current flow.

Series/Parallel Circuit A circuit with both series and parallel branches.

Service Bulletin A publication that provides the latest service tips, field repairs, product improvements, and related information of benefit to service personnel.

Service Manual A manual, published by the manufacturer, that contains service and repair information for all vehicle systems and components.

Shift Fork The Y-shaped component located between the gears on the main shaft that, when actuated, causes the gears to engage or disengage via the sliding clutches.

Shift Rail Shift rails guide the shift forks using a series of grooves, tension balls, and springs to hold the shift forks in gear.

Shift Tower The main interface between the driver and the transmission, consisting of a gearshift lever, pivot pin, spring, boot and housing.

Shift Yoke A Y-shaped component located between the gears on the main shaft that, when actuated, causes the gears to engage or disengage via the sliding clutches.

Shock Absorber A hydraulic device used to dampen vehicle spring oscillations for controlling body sway and wheel bounce, and/or prevent spring breakage.

Short Circuit An undesirable connection between two worn or damaged wires. The short occurs when the insulation is worn between two adjacent wires and the metal in each wire contacts the other, or when the wires are damaged or pinched.

Single-Axle Suspension A suspension with one axle.

Single-Reduction Axle Any axle assembly that uses a single gear reduction in the differential carrier.

Slide Travel The distance that a sliding fifth wheel is designed to move.

Sliding Fifth Wheel A fifth wheel design that slides fore and aft to adjust weight distribution on the tractor axles and/or overall length of the tractor and trailer.

Slipout A condition that generally occurs when pulling with full power or decelerating with the load pushing. Tapered or worn clutching teeth will try to "walk" apart as the gears rotate, causing the sliding clutch and gear to slip out of engagement.

Slip Rings and Brushes Components of an alternator that conducts current to the rotor. Most alternators have two slip rings mounted directly on the rotor shaft; they are insulated from the shaft and from each other. A spring loaded carbon brush is located on each slip ring to carry the current to and from the rotor windings.

Solenoid An electromagnet that is used to effect linear or rotary movement.

Solid-State Device A semiconductor device.

Solid Wires A single-strand conductor.

Solvent A substance that dissolves other substances.

Spade Fuse A term used for blade fuse.

Spalling Surface fatigue that occurs when chips, scales, or flakes of metal break off due to fatigue rather than wear. Spalling is usually found on splines and U-joint bearings.

Specialty Service Shop A shop that specializes in areas such as engine rebuilding, transmission/axle overhauling, brake, air conditioning/heating repairs, and electrical/electronic work.

Specific Gravity The ratio of a liquid's mass to an equal volume of distilled water.

Spiral-Bevel Gear A gear arrangement that has a drive pinion gear that meshes with the ring gear at the centerline axis of the ring gear. This gearing provides strength and allows for quiet operation.

Splined Yoke A yoke that allows the driveshaft to increase in length to accommodate movements of the drive axles.

Spontaneous Combustion A process by which a combustible material self-ignites.

Spread-Tandem Suspension A two-axle assembly in which the axles are spaced to allow maximum axle loads under existing regulations. The distance is usually more than 55 inches.

Spring A device used to reduce road shock and transfer loads through suspension components to the frame of the truck or trailer.

Spring Chair A suspension component used to support and locate the spring on an axle.

Spring Deflection The depression of a suspension when the springs are placed under load.

Spring Rate The load required to deflect the spring a given distance, (usually one inch).

Spring Spacer A riser block often used on top of the spring seat to obtain increased mounting height.

Stability A relative measure of the handling characteristics that provide the desired and safe operation of the vehicle during various maneuvers.

Stabilizer A device used to stabilize a vehicle during turns; sometimes referred to as a sway bar.

Stabilizer Bar A bar that connects the two sides of a suspension so that cornering forces on one wheel are shaped by the other. This helps equalize wheel side loading and reduces the tendency of the vehicle body to roll outward in a turn.

Stand Pipe A type of check valve that prevents reverse flow of the hot liquid lubricant generated during operation. When a universal joint is at rest, one or more of the cross ends will be up. Without the stand pipe, lubricant would flow out of the upper passage ways and trunnions, leading to partially dry startup.

Starter Circuit The circuit that supplies power for the engine cranking.

Starter Motor The device that converts the electrical energy from the battery into mechanical energy for cranking the engine.

Starting Safety Switch A switch that prevents vehicles with automatic transmissions from being started in gear.

Stationary Fifth Wheel A fifth wheel whose location on the tractor frame is fixed once it is installed.

Stator A component located between the pump/impeller and turbine to redirect the oil flow from the turbine back into the impeller in the direction of impeller rotation with minimal loss of speed or force.

Stator Assembly The torque-converter reaction member or torque multiplier supported on a free wheel roller race splined to the front support assembly.

Steering Gear The steering-box assembly mounted in a housing that converts steering-wheel motion into axle steer.

Steering Stabilizer A shock absorber attached to the steering components to cushion road shock in the steering system, improving driver control in rough terrain and protecting the system.

Stepped Resistor A resistor designed to have two or more fixed values, by connecting wires to one of several taps.

Stoplight Switch A pneumatic switch that actuates the brake lights. There are two types: (1) A service stoplight switch that is located in the service circuit, actuated when the service brakes are applied. (2) An emergency stoplight switch located in the emergency circuit and actuated when a pressure loss occurs.

Storage Battery A lead-acid battery.

Structural Member A primary load-bearing portion of the body structure that affects its over-the-road performance or crash-worthiness.

Sulfation A condition that occurs when sulfate is allowed to remain in lead-acid battery plates for a long time, usually resulting in a destroyed battery.

Suspension The system that connects the axles to the vehicle frame, designed so that road shock is dampened.

Suspension Height The distance from a specified point on a vehicle to the road surface.

Swage To reduce or taper.

Sway Bar A component that connects the two sides of a suspension so that cornering forces on one wheel are shared by the other. This helps equalize wheel side loading and reduces the tendency of the vehicle body to roll outward in a turn.

Switch A device used to control on/off and direct the flow of current in a circuit. A switch can be under the control of the driver or can be self-operating.

Synchromesh A mechanism that equalizes the speed of the gears that are clutched together.

Synchro-transmission A transmission with mechanisms for synchronizing the gear speeds so that the gears can be shifted without clashing, thus eliminating the need for double-clutching.

System Protection Valve A valve to protect the brake system against an accidental loss of air pressure, buildup of excess pressure, or back-flow and reverse airflow.

Tachometer An instrument that indicates rotating speeds, sometimes used to indicate crankshaft rpm.

Tag Axle A nondrive rearmost axle of a tandem axle tractor used to increase the load-carrying capacity of the vehicle.

Tapped Resistor A resistor designed to have two or more fixed values, available by connecting wires to either of the several taps.

Tandem Axle Suspension A suspension system consisting of two axles with a means for equalizing weight between them.

Tandem Drive A two-axle drive combination.

Tandem Drive Axle Double axle assembly using an interaxle differential divider and a shaft that connects the two axle carriers.

Three-Speed Differential A type of axle in a tandem two-speed axle arrangement with the capability of operating the two drive axles in different speed ranges at the same time.

Throw (1) The offset of a crankshaft. (2) The number of output circuits of a switch.

Tie-Rod Assembly A shaft that links the two steering knuckles on either side of an axle and forces them to act in unison.

Time Guide Prepared reference material used for computing compensation payable by the truck manufacturer for repairs or service work to vehicles under warranty.

Timing (1) A procedure of marking the teeth of a gear set prior to installation and placing them in proper mesh. (2) Spark delivery in relation to the piston position.

TMC The Maintenance Council of the American Trucking Association. Sets standards for the trucking industry including that for wheel-end procedure.

Toe A suspension dimension that reflects the difference in the distance between the extreme front and rear of the tire.

Toe-In A suspension dimension whereby the front of the tire points inward toward the vehicle.

Toe-Out A suspension dimension whereby the front of the tire points outward from the vehicle.

Top U-Bolt Plate A plate located on the top of the spring and is held in place when the U-bolts are tightened to clamp the spring and axle together.

Torque To tighten a fastener to a specific degree of tightness, generally in a given order or pattern if multiple fasteners are involved on a single component.

Torque and Twist A term that generally refers to the forces developed in the trailer and/or tractor frame that are transmitted through the fifth wheel when a rigid trailer, such as a tanker, is required to negotiate bumps, like street curbs.

Torque Converter A component device, similar to a fluid coupling, that transfers engine torque to the transmission input shaft and can multiply engine torque by having one or more stators between the members.

Torque Rod Shim A thin wedge-like insert that rotates the axle pinion to change the U-joint operating angle.

Torsional Rigidity A component's ability to remain rigid when subjected to twisting forces.

Total Pedal Travel The complete distance the clutch or brake pedal must move.

Toxicity A measure of how poisonous a substance is.

Tracking The travel of the rear wheels of a vehicle in relation to the front wheels.

Tractor A motor vehicle, without a body, that has a fifth wheel and is used for pulling a semitrailer.

Tractor Protection Valve A device that automatically seals off the tractor air supply from the trailer air supply when system pressure drops to 20 or 45 psi.

Tractor/Trailer Lift Suspension A single axle air ride suspension with lift capabilities commonly used with steerable axles for pusher and tag applications.

Trailer A platform or container on wheels pulled by a car, truck, or tractor.

Trailer Hand-Control Valve A device located on the dash or steering column and used to apply only the trailer brakes.

Trailer Slider A movable trailer suspension frame that is capable of changing trailer wheelbase by sliding and locking into different positions.

Transfer Case An additional gearbox located between the main transmission and the rear axle to split torque from the transmission between front and rear driving axles.

Transistor A three terminal, solid-state device used as an electronic relay or amplifier.

Transmission A device used to transmit engine torque at various ratios.

Transverse Vibrations A condition caused by an unbalanced driveline or bending movements, in the driveshaft.

Treadle A dual-brake foot valve that manages truck-service brake applications.

Treadle Valve A foot-operated brake valve that controls air pressure to the service chambers.

Tree Diagnosis Chart A chart used to provide a logical sequence for what should be inspected or tested when troubleshooting a repair problem.

Triaxle Suspension A suspension consisting of three axles with a means of equalizing weight between axles.

Trunnion The end of the universal cross; they are case hardened ground surfaces on which the needle bearings ride.

TTMA Acronym for Truck and Trailer Manufacturers Association.

Turbine The output (driven) member of a torque converter that is splined to the forward clutch of the transmission and to the turbine shaft assembly.

TVW Acronym for (1) Total vehicle weight. (2) Towed vehicle weight.

Two-Speed Axle Assembly An axle assembly having two different output ratios from the differential.

U-Bolt A fastener used to clamp the top U-bolt plate, spring, axle, and bottom U-bolt plate together.

Underslung Suspension A suspension in which the spring is positioned under the axle.

Universal Gladhand A term often used for nonpolarized gladhand.

Universal Joint (U-joint) A shaft component that allows torque to be transmitted at different angles.

Upper Coupler The flat load-bearing surface under the front of a semitrailer, including the kingpin, which rests on the fifth wheel when coupled.

Vacuum Pressure valves below atmospheric pressure.

Validity List A list supplied by the manufacturer of valid bulletins.

Valve Body and Governor Test Stand Automatic transmission test equipment. The valve body of the transmission is removed from the vehicle and mounted into the test stand. The test stand duplicates all vehicle running conditions, so the valve body can be calibrated.

Variable-Pitch Stator A stator design often used in torque converters in off-highway applications.

Vehicle Body Clearance (VBC) The distance from the inside of the inner tire to the spring or other body structures.

Vehicle On-board Radar (VORAD) A Dopler-radar collision warning system.

Vehicle Retarder An auxillary braking device designed to supplement the service brakes on heavy-duty trucks.

Vertical Load Capacity The maximum, recommended vertical downward force that can be safely applied to a coupling device.

VIN Acronym for Vehicle Identification Number.

Viscosity A measure of the flow and sheer resistance of a lubricating oil.

Volt The unit of electromotive force.

Voltage Regulator A device that controls the voltage in the charging circuit.

VORAD An acronym for Vehicle On-board Radar.

Vortex Oil Flow The circular flow that occurs as the oil is forced from the impeller to the turbine and then back to the impeller.

Watt The measure of electrical power.

Wedge-Actuated Brakes A brake system using air pressure and air brake chambers to drive a wedge and roller assembly into an actuator located between brake shoes.

Wet Tank A supply reservoir.

Wheel Alignment The mechanics of keeping all the parts of the steering system in the specified relation to each other.

Wheel- and Axle-Speed Sensors Electromagnetic devices used to signal wheel speed data.

Wheel Balance The equal distribution of weight in a wheel with the tire mounted. It is an important factor that affects tire wear and vehicle control.

Windings (1) The three separate bundles in which wires are grouped in the stator. (2) The coil of wire found in a relay or other similar device. (3) That part of an electrical clutch that provides a magnetic field.

Work (1) Forcing a current through a resistance. (2) The product of a force.

Yield Strength The highest stress a material can stand without permanent deformation or damage, expressed in pounds per square inch (psi).

Zener Diode A variation of the diode, this device functions like a standard diode until a certain voltage is reached. When the voltage level reaches this point, the zener diode will allow current to flow in the reverse direction. Zener diodes are often used in electronic voltage regulators.